The ARRL Satellite Anthology

Published by
The American Radio Relay League
225 Main Street,
Newington, Connecticut 06111

THE COVER

A—Photo courtesy of Grant Zehr, WA9TFB

B—Photo of AMSAT-OSCAR 10 courtesy AMSAT-DL

C—Photo of UoSAT-OSCAR 11 courtesy of the University of Surrey, England

D—Photo courtesy of SRI International, Menlo Park, California

Copyright © 1988 by
The American Radio Relay League

Copyright secured under the Pan-American Convention

International Copyright secured

This work is publication No. 87 of the Radio Amateur's Library, published by the League. All rights reserved. No part of this work may be reproduced in any form except by written permission of the publisher. All rights of translation are reserved.

Printed in USA

Quedan reservados todos los derechos

First Edition

$5.00 in USA and Possessions
$5.50 in Canada and elsewhere, US Funds

ISBN: 0-87259-210-3

Foreword

OSCAR (Orbiting Satellite Carrying Amateur Radio) has been a part of our hobby since the early 1960s. OSCAR I was launched on December 12, 1961. OSCAR III, the first amateur satellite with a "transponder" was launched on March 9, 1965. Operating through the early satellites was a complicated venture at best. In the almost 30 years since the launch of OSCAR I, amateur satellite technology has seen great advances, contributing to the exponential growth in the popularity of OSCAR operation. The successful launch of the Phase 3C satellite (now known as OSCAR 13) provides another addition to the growing number of active satellites. In this book, we have collected a range of articles from QST that will tell you enough about the satellites that are in operation today (and a few that are planned for tomorrow) to enable you to join in the fun and excitement of this fascinating and enjoyable part of ham radio. The satellites are waiting; I hope to hear you soon!

David Sumner, K1ZZ
Executive Vice President, ARRL

Preface

Amateur Radio has experienced many significant trends over its proud history. Whether the transition is from from regenerative to superheterodyne receivers or from tube to transistor amplifiers, the tides of change periodically sweep over Amateur Radio. When the wave has passed, the hobby is left refreshed; renewed; alive with a new direction and vitality to advance to the next major wave of change.

Today we see the confluence of *two* major waves of change: the digital revolution and the maturing of OSCAR satellites. These waves of change may be the most significant in our history. The digital revolution promises to change the way we communicate, while satellite communications unshackle us from dependence on the ionosphere for over-the-horizon DX communications. Moreover, OSCAR allows us to use the vast spectral resources at VHF and above for a host of activities beyond line-of-sight.

In significant ways, the advance of digital communications and the advances in satellite communications are related. For example, computers are at the heart of packet-radio communications, and computers also help make OSCAR satellite use more fun and notably easier. With relatively inexpensive personal computers in many ham shacks, the effort of tracking satellites has been reduced from a difficult chore to a fascinating, simple task. Digital communications will soon affect all aspects of our hobby. Even our voice communications may be digitized on the ham bands soon; digital audio communications are already in use on many telephone circuits we use daily. And the digital revolution will give us time-independent communication if we choose. We can leave messages in convenient, short-term repositories and pick up replies there later as part of a satellite-borne electronic mail system.

With AMSAT OSCAR 13 representing a family of high-orbiting, long duration satellites, the style of satellite QSOs is changing. Now, leisurely rag chews and intercontinental round table discussions are easy. This is a far cry from earlier satellites where short-duration passes and QRN were the rule.

With advanced modes of operation on UHF, excellent quality signals can be realized with surprisingly modest rigs. The rigs themselves are now highly digital. Frequency synthesis, digital memory and similar features make working OSCAR easier and more rewarding than ever before. Advanced OSCAR stations even have automatic antenna tracking to keep the satellite in view.

Using digital techniques, telemetry capture and processing can be automated, and the telemetry reveals much of the inner workings of the satellite. This too is a far cry from the days when legions of dedicated trackers copied reams of telemetry reports sent in slow CW.

If, as they say, proliferation of a species is a sign of a healthy breed, then the OSCAR "bird" must be vivacious indeed! AO-13 was launched in June 1988. Even while it was on its way to orbit, new satellites were being designed and built in several shops around the world. These will lead to a bumper crop of new satellites in the closing years of the Eighties. These new satellites will include advanced digital communications satellites, educational satellites providing easy access for students and combinations of traditional communications missions.

The collection of articles you now hold provides a glimpse of what's in store for today's OSCAR user. With this digest in hand, you have opened the door to tomorrow. But be forewarned! Once you experience OSCAR, you may never be happy with the plain old ionosphere again, just as having twirled the knob on your new, all-mode digitally synthesized rig, you will never again be content with that vintage NC-300!

Now... let the *real* adventure begin!

Vern "Rip" Riportella, WA2LQQ
AMSAT President
June 1988

Contents

Introducing Phase 3C: A New, More Versatile OSCAR
by Vern Riportella, WA2LQQ 1

Adventures In Satellite DXing—Part 1
by Dick Jansson, WD4FAB 10

Adventures In Satellite DXing—Part 2
by Dick Jansson, WD4FAB and Mark Wilson, AA2Z 13

Adventures In Satellite DXing—Part 3
by Dick Jansson, WD4FAB 19

Adventures In Satellite DXing—Part 4
by Dick Jansson, WD4FAB and Mark Wilson, AA2Z 26

OSCAR at 25: The Amateur Space Program Comes of Age
by Jan King, W3GEY, Vern Riportella, WA2LQQ and Ralph Wallio, W0RPK 33

OSCAR at 25: Beginning of a New Era
by Jan King, W3GEY, Vern Riportella, WA2LQQ and Ralph Wallio, W0RPK 37

A Mode-L Parabolic Antenna and Feedhorn for OSCAR 10
by Eugene F. Ruperto, W3KH 42

Microcomputer Processing of UoSat-OSCAR 9 Telemetry
by Robert J. Diersing, N5AHD 46

A Profile of the UoSat-OSCAR 11 Satellite
by Jon Bloom, KE3Z 52

Amateur Satellite Communications
by Vern Riportella, WA2LQQ

General Operating

Antennas for Working OSCAR 56
Basic Satellite Tracking Themes 57
Basic Satellite Tracking Themes—Part 2 58
Fun, Games and (Hopefully) Technical Challenges on OSCARs 59
Fun, Games and (Hopefully) Education on OSCARs—Part 2 61
Fun, Games and (Hopefully) Education on OSCARs—Part 3 63
Fun, Games and (Hopefully) Education on OSCARs—Part 4 64
Prospects for Mobile and Portable OSCAR Operation 65
Waves in Rotation: The Challenge of Circular Polarization 66
Where to Get OSCAR Information 67
Working OSCAR—The Basics 68

Copyright © 1988 by

The American Radio Relay League

Copyright secured under the Pan-American Convention

International Copyright secured

This work is publication No. 87 of the Radio Amateur's Library, published by the League. All rights reserved. No part of this work may be reproduced in any form except by written permission of the publisher. All rights of translation are reserved.

Printed in USA

Quedan reservados todos los derechos

First Edition

$5.00 in USA and Possessions
$5.50 in Canada and elsewhere, US Funds

ISBN: 0-87259-210-3

The ARRL Satellite Anthology

Published by
The American Radio Relay League
225 Main Street,
Newington, Connecticut 06111

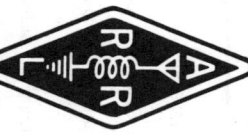

THE COVER

A—Photo courtesy of Grant Zehr, WA9TFB
B—Photo of AMSAT-OSCAR 10 courtesy AMSAT-DL
C—Photo of UoSAT-OSCAR 11 courtesy of the University of Surrey, England
D—Photo courtesy of SRI International, Menlo Park, California

Index

Page numbers in *italics* refer to entries in the fact box, on the map, or in the glossary.

Aborigines 5, 6, 8, 26, 30
Adelaide *4, 5,* 20
Alice Springs *5,* 8
amusement parks 29
animals 6, 8, 30
Aussie Rules 16
Ayers Rock *See* Uluru

barbecues 24, 28
bauxite 15
beaches 8, 9, 17, 28
bicycles 26, 29
Brisbane *4, 5,* 20
buses 26

camping 28
Canberra *4, 5,* 6, 7
cars 20
cattle 14, 21
cattle stations 12
cave paintings 8
cell phones 11, 23
children 23, 26, 27
Christianity *4*
cities 4, 6, 7, 10, 11, 20, 29
coal 15
computers 23
continent *4, 31*
coral reefs *5, 31*

crest *6, 31*
crops 14, 19
currency *4,* 19, *31*

Darwin *4, 5*
desert *4,* 8
diamonds 15

families 20, 22, 23, 29
farmers 12, 14
fire 30
fish 9, 24, 25
Flying Doctor Service 13, 21
food 13, 18, 19, 24, 25
fruit 14, 18

gold 15
government 6
grapes 15
Great Barrier Reef *5,* 9

Hobart *4, 5*
homesteads 12
homes 10, 12, 22, 23

lakes 7
life-saving clubs 17, 28

markets 14, 18
marsupial 8, *31*
meat 14, 24, 25
Melbourne *4, 5,* 6, 20
mining 14, 15

monorail 20, *31*
mountains *4,* 5

national parks 8, 28
New South Wales *5*
Northern Territory *5,* 8, 14

Olympic Games 16
opals 15
outback 12, 13, 14, 15, 21, 27

Parliament buildings 6
Perth *4, 5*
police 10

Queensland *4, 5,* 14

rain forest *4, 31*
refinery 14, *31*
restaurants 11, 24, 25
rivers *4, 5,* 7
road trains 21
roads 20, 21
runways 21

School of the Air 27
schools 26, 27
sheep 14, 21
sheep shearing 12
shops 11, 18, 19
silver 15
South Australia *5*
sports 9, 16, 17, 22, 26, 28, 29

state *4, 31*
suburbs 10, 20, *31*
surfing 9, 17
Sydney *4, 5,* 16, 20, 24
Sydney Harbor 9
Sydney Opera House 9

Tasmania *4, 5*
television 22
territory *4, 31*
tourists 7, 9
towns 10, 12, 13, 14, 18
traffic 20, 21
trains 20

Uluru *5,* 8
ute 13

Victoria *5*

water 13
Western Australia *5*
wine 15
wool 14

32

Glossary

Artesian water Water that has been taken from underground.

Commuter Someone who travels some distance from home to work each day.

Continent One of the large landmasses of the world.

Coral reef A bank of coral just below sea level.

Crest An official badge.

Currency The money used in a country.

Exporter Someone who sells a product from one country to another.

Marsupial A mammal that carries its young in a pouch.

Monorail A railroad with only one rail.

Nocturnal Active at night.

Ore A solid, rocklike material that contains a valuable metal.

Population The number of people who live in one place.

Rain forest A lush, dense forest found in hot regions with high rainfall.

Refinery A place where raw materials are turned into a product.

Smelt To melt ore to extract metal from it.

State A region of a country that has its own government for local affairs.

Suburbs Housing on the outer areas of a city.

Territory A region of a country that has its own government for local affairs, like a state.

Verandah A covered space along the side of a building.

Going further

Aboriginal painting
Lots of Aboriginal paintings are made by using the end of a stick instead of a brush.

Look closely at the boomerang and use the same painting technique to make your own pictures and designs.

Candy wrappers
The Australian candy bars below each feature a different Australian animal.

Design some candy wrappers, using a different Australian theme, for example, sports or famous sights.

A fire poster
Fire is one of the greatest dangers to life and property in Australia.

Design a poster to encourage people not to light fires when the weather is very hot and dry.

Websites
www.about-australia.com
www.frogandtoad.com.au
www.auslig.gov.au

Having fun

△ **Cycling**
Families use mountain bikes to explore the rugged landscape.

◁ **Basketball**
Even children in remote outback communities may have a basketball court.

△ **Amusement parks**
People enjoy daredevil rides in city amusement parks.

Amateur Satellite Communications continued

Phase 3 (OSCAR 10 and OSCAR 13)
The Digital Satellite World 70
Gateways: Keys to Opening New Communications Doors 72
Introducing Phase 3C: Newest High Flyer Debuts Soon 74
Introducing Phase 3C: "Superbird" Soon! 76
Phase 3C: What Do You Need To Work It? 78
Phase 3C Operating Schedules 80
RUDAK—What Is It? 82
The AMSAT RUDAK User Terminals 84

JAS-1 (Fuji-OSCAR 12)
Birth of a New OSCAR: First All-Japanese Project Debuts 86
Introducing Japanese Amateur Satellite Number One (JAS-1) 88
Operating the Flying Mailbox: FO-12 Mode JD 90

Soviet Spacecraft (RS-10/11)
New Russian Satellite Sparks Surge of Interest 92
New Russian Satellite Sparks Interest Surge—Part 2 93
New Russian Satellite Sparks Interest Surge—Part 3 95

Phase 4
The Future Is Up There 96
NEXT Generation Satellites on the Horizon 97

Introducing Phase 3C: A New, More Versatile OSCAR

Easily accessible and easy to use, Phase 3C offers many operating alternatives to those of previous Amateur Radio satellites.

By Vern Riportella, WA2LQQ
PO Box 177
Warwick, NY 10990

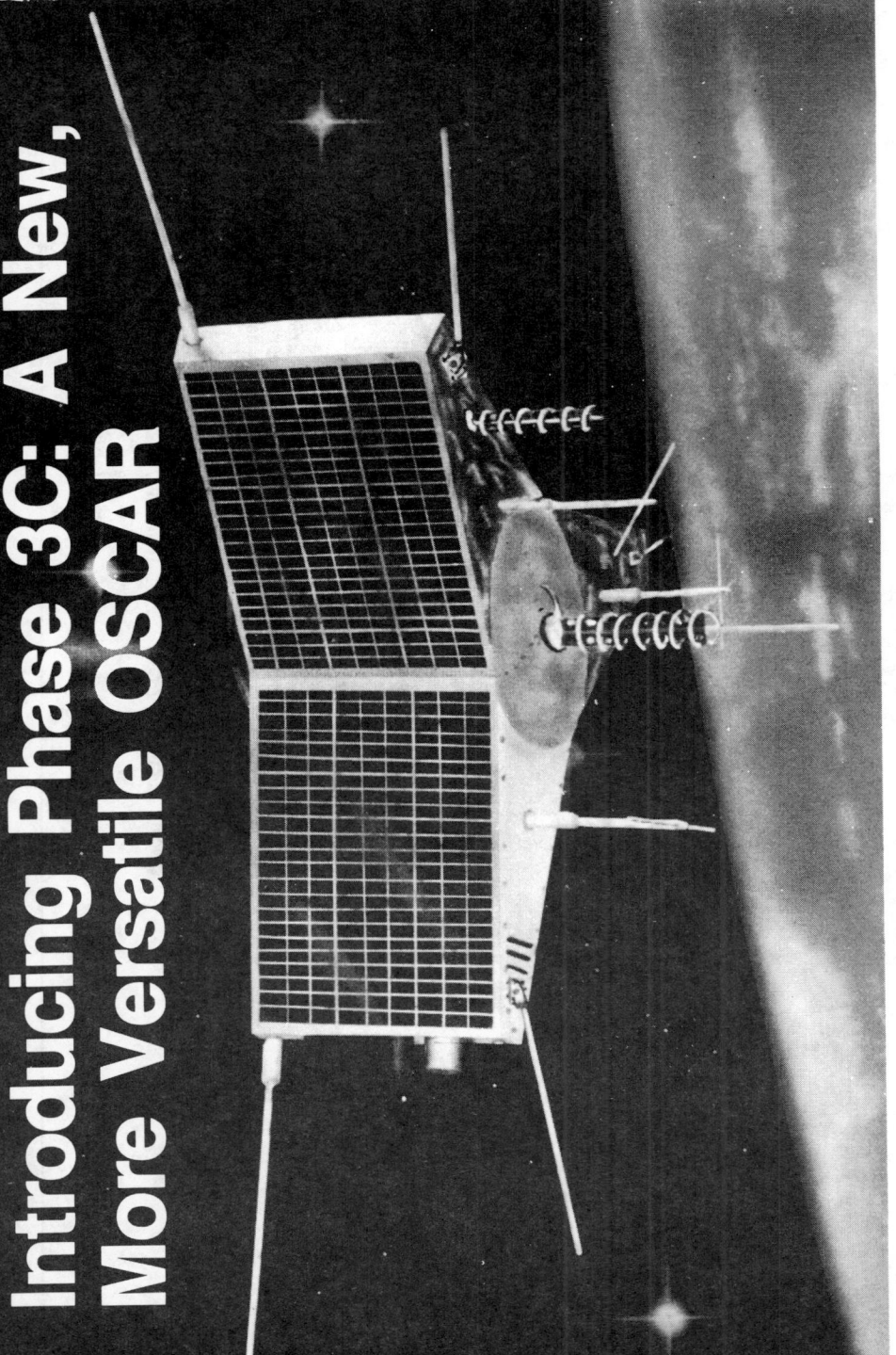

What has been around for more than 25 years, has been used by tens of thousands of hams, including a US President and a king, and now stands poised for a new series of breakthroughs? Since this is an article about OSCAR, you probably guessed the answer right away![1]

But suppose I made the question slightly more difficult. Did you know the latest OSCAR is ready for launch and will permit intercontinental QSOs on voice, CW, SSTV or packet?

It's a fact. As you read this, the newest OSCAR may have already been placed in orbit. A new Amateur Radio satellite will be plying its way across the heavens, providing superb VHF and UHF communications to those who know how to use it. Surprisingly, using the new OSCAR will be easier than ever—even though it's by far the most complex and versatile OSCAR ever launched.

OSCARs Are Easier Than Ever To Use

The new ease of operation results from several significant factors, including noteworthy improvements in the area of user support. For example, tracking OSCARs used to be a chore. Now, computer programs make tracking a breeze. Some of the newer systems automatically point your antennas. And if your radios can be interfaced to a computer, there are programs to tune your radios to compensate for Doppler shift. You just sit back and enjoy the QSOs!

Furthermore, an improved network of OSCAR Elmers is being established to ensure you get the help you need when getting started. The equipment needed to work any OSCAR is readily available. Many quality VHF and UHF transceivers can be found on the new- and used-equipment markets. (I still use a vintage TS-700S I purchased 12 years ago for OSCAR work.) Antennas and preamps proliferate in magazine ads and on flea-market tables. Moreover, OSCAR users are a resourceful group and often enjoy renewing the "home-brew" legacy in their own shacks. Many homemade helical antennas are regularly used for OSCAR operation. In sum, Phase 3C will be easier to use. Taking advantage of technological advances, the new OSCAR will provide more payback in user satisfaction than any of the earlier amateur satellites.

What is Phase 3C?

Phase 1 OSCARs were simple battery-powered RF beacons usually placed in low orbits (the first was launched in 1961). They had typical life spans of only a few weeks to a few months. Phase 2 OSCARs were more sophisticated. They carried solar cells and batteries for longer life spans. Most important, Phase 2 OSCARs carried repeaters, or, as they are properly called, *transponders*. Phase 2 satellites such as AMSAT-OSCARs 6, 7 and 8 provided useful VHF/UHF communication between amateurs separated 3000 miles and, on occasion, more. Using AMSAT-OSCAR 7, contacts between the East Coast and

Table 1
Phase 3C Operating Modes and Frequencies

Mode	Uplink Band	Downlink Band	Notes
B	70 cm (435 MHz)	2 m (145 MHz)	AO-7 and AO-10 favorite
S	70 cm (435 MHz)	13 cm (2401 MHz)	New; first use on Phase 3C
JL	2 m & 24 cm (1269 MHz)	70 cm (435 MHz)	New; first use on Phase 3C
RUDAK	24 cm (1269 MHz)	70 cm (435 MHz)	New; first use on Phase 3C

Beacons Mode B: General beacon 145.812; Engineering beacon 145.985 MHz
Mode JL: General beacon 435.651; Engineering beacon/RUDAK downlink : 435.677 MHz
Mode S: Beacon: 2400.325 MHz

Mode B

Uplink		Downlink	
435.420		145.985 MHz	Engineering beacon
435.430		145.975	Passband limit, upper
435.440		145.965	
435.450		145.955	
435.460		145.945	
435.470		145.935	
435.475		145.925	
435.480		145.915	
435.490		145.905	
435.495		145.895	
435.500		145.890	Passband center
435.505		145.885	
435.510		145.875	
435.520		145.865	
435.530		145.855	
435.540		145.845	
435.550		145.835	
435.560		145.825	
435.570		145.812	General beacon

Mode JL

L Uplink	J Uplink	Downlink	
1269.330		436.005 MHz	Passband limit, upper
1269.340		435.995	
1269.345	144.425	435.990	J sub-band limit, upper
1269.355	144.435	435.980	
1269.365	144.445	435.970	
1269.375	144.455	435.960	
1269.385	144.465	435.950	
1269.395	144.475	435.940	J sub-band limit, lower
1269.400		435.935	
1269.410		435.925	
1269.420		435.915	
1269.430		435.905	
1269.440		435.895	
1269.450		435.885	
1269.460		435.875	
1269.470		435.865	
1269.475		435.860	Passband center
1269.480		435.855	
1269.490		435.845	
1269.500		435.835	
1269.510		435.825	
1269.520		435.815	
1269.530		435.805	
1269.540		435.795	
1269.550		435.785	
1269.560		435.775	
1269.570		435.765	
1269.580		435.755	
1269.590		435.745	
1269.600		435.735	
1269.610		435.725	
1269.620		435.715	Passband limit, lower
1269.710	RUDAK up	435.677	Engineering beacon/RUDAK downlink
		435.675	
		435.665	
		435.655	
		435.651	General beacon

Mode S

Uplink	Downlink	
435.601	2400.325 MHz	Beacon
435.605	2400.711 MHz	Passband limit, lower
435.610	2400.715	
435.615	2400.720	
435.619	2400.725	
435.620	2400.729	
435.625	2400.730	Passband center
435.630	2400.735	
435.635	2400.740	
435.637	2400.745	Passband limit, upper
	2400.747	

RUDAK

Uplink	Downlink	
1269.710 MHz	435.677 MHz	Single channel

Although this table might suggest the passbands are divided into many 10-kHz channels, this is not the case; there are no "channels" as such. The table merely shows the correlation of uplink and downlink frequencies across a continuous spectrum of available frequencies with certain "guideposts" illustrated for convenience.

Hawaii could be accomplished under some conditions.[2]

By 1976, however, designers sought to provide regular intercontinental linkups and longer periods of coverage. They recognized that aside from the recreational aspects of Amateur Radio, OSCARs could play a vital role in public-service emergency communications. But, emergency communicators cannot wait for a satellite to appear; a communications channel has to be available nearly always. Recognizing this fundamental public-service requirement, together with the existing Amateur Radio community's appetite for longer access and ease of use, the OSCAR designers began to raise their sights above the low earth-orbiting (LEO) Phase 2 OSCARs.

At first, the designers considered a geosynchronous satellite, but soon realized the inherent limits. For one thing, it would be expensive, and only one could be built. Because it would be expensive, it had to be the product of a large consortium of Amateur Radio satellite groups. But, if the satellite were geosynchronous, its communications area would cover less than one third of the earth's surface. Some countries would be left out. How could the geosynchronous satellite be positioned to serve the contributors equitably?

The answer is the series of Phase 3 satellites. By using a type of orbit pioneered by the Soviets called *Molniya*, a kind of semi-synchronous orbit could be achieved. The *Molniya* orbit is elliptical instead of circular, and has a period of about 12 hours. For much of the orbit the satellite remains high above the earth (nearly 36,000 km;

22,000 miles) to provide a footprint as large as that of a geosynchronous satellite. Then the bird plummets close to earth only to rise toward its apogee (high point) elsewhere. In this way, Phase 3C provides benefits of broad coverage that can be distributed among a worldwide user population. In other words, the Phase 3 Molniya orbit gives the next best thing to intercontinental coverage all of the time: intercontinental coverage *much* of the time. The Molniya orbit spreads that coverage equitably across the globe, obviating difficult decisions on positioning a geosynchronous satellite's footprint.

By 1980, the first Phase 3 bird, Phase 3A, sat on the launch pad in French Guiana. From there, it was to fly into a Molniya orbit from which it could serve a worldwide community of OSCAR users clamoring for the fun of satellite DX. Many looked forward to working rare DX without regard to the vagaries of F_2 propagation. Others sought to make obsolete the hurried QSOs so common on the low-orbiting Phase 2 satellites that quickly whizzed from horizon to horizon.

The Phase 3A project ended in disaster, however, when an errant launcher was destroyed, dropping its payload into the Atlantic ocean. The product of years of work—transponders, computers, everything—literally went down to the sea in chips!

Not to be bested by "outrageous fortune," the AMSAT team went back to work to develop Phase 3B in 1983. Phase 3B became AMSAT-OSCAR 10 upon its successful insertion into orbit. Since then, nearly 150 countries have been heard on AO-10, and numerous firsts have been achieved. Now, the stage is set for the latest, most powerful and capable OSCAR yet: Phase 3C, the third in its generation.

Phase 3C Subsystems

Phase 3C is a self-contained system. It carries all the resources it will need to operate in space for 5 years or more. The satellite's four major subsystems are: communications, power, control and propulsion.

The Communications Subsystem

The Phase 3C communications subsystem consists of four transponders. Each of these functions as a cross-band repeater. (See Table 1.) You transmit on the uplink frequency and listen on the downlink frequency. Because you transmit and receive on different bands, you can transmit and receive simultaneously (full duplex operation), and even listen to your downlink signal delayed up to a third of a second—an experience that is unnerving to those who've not experienced it before!

The Mode B and JL transponders are linear. That means the signal appearing in the uplink receiver is transmitted on the downlink so that SSB, CW, SSTV and various digital waveforms are faithfully reproduced. Moreover, the transponder passbands are wide enough to accommodate several contacts at once. That means European amateurs can chat with those in South America, while hams in Africa can chat with North Americans and others in between at the same time, albeit on different frequencies within the passband.

The Mode-S transponder is an experimental UHF-to-UHF transponder designed primarily for a single FM channel. It can handle up to four SSB stations, providing they each stay below the hard-limiting threshold of the transponder. RUDAK—the English translation: *Regenerative Repeater for Digital Amateur Communication*—is a new type of packet-radio repeater. It uses the most advanced form of digital modulation (phase shift keying, PSK) and techniques to create an efficient digital communications channel.

The antennas on Phase 3C are an assortment of VHF and UHF designs. Coverage of the 2-meter band is accomplished using two antennas. The 2-meter high-gain array, a so-called ZL Special, consists of three two-element beams phased in such a way as to produce right-hand circular (RHC) polarization.[3] This array has a gain of about 6 dBic (decibels referenced to a circularly polarized isotropic source), and a 3-dB beamwidth of roughly 100°. An omnidirectional 2-meter antenna is used for conditions where the high-gain beam is inappropriate. This antenna is linearly polarized, has as much as 2 dBi gain and a toroidal pattern.

The UHF antennas on Phase 3C are common types. On 70 cm, a monopole is also used for the same reasons as the 2-meter monopole, and it gives comparable performance. The 70-cm high-gain array consists of three phased dipoles over a groundplane, and produces 9.5 dBic gain with a theoretical 3-dB beamwidth of 67°. The 24-cm antenna is a 5-turn RHC helix supported by a frame. Its gain is 12.2 dBic and its 3-dB beamwidth is 49°. The 13-cm antenna is a 6-turn RHC helix providing 13 dBic gain and having a 3-dB beamwidth of 45°.

The satellite transmits a strong signal that can be heard well across a distance of as much as 40,000 km (25,000 miles). Phase 3C's effective radiated power referenced to an isotropic source (EIRP) is shown in Table 2.

The Power Subsystem

The Phase 3C power subsystem consists of the solar-cell arrays, battery charge regulator and batteries. The solar arrays are made up of hundreds of silicon solar cells on six separate surfaces. Together, they generate 50 watts when first placed in orbit.

Table 2
Phase 3C Effective Radiated Power, Isotropic (EIRP)

Mode	Band	EIRP PEP	EIRP Average	Peak/Average
B	2 m	23 dBW (200 W)	17 dBW (50 W)	6 dB
JL	70 cm	26.5 dBW (446 W)	20.5 dBW (111 W)	6 dB
S	13 cm	0.97 dBW (1.25 W)	(continuous)	0 dB

View of one arm of Phase 3C showing the sun sensor and earth sensor (left), sensor electronics unit (center) and propulsion flow assembly (right). *(photo by Robert J. Diersing, N5AHD)*

Satellite Anthology 3

Table 3
Phase 3C Preliminary Telemetry Channel Allocations and Estimated Equations

Telemetry Channel	Function	Typical Equation (Subject to final cal.)	Units
00	Solar panel out and BCR input voltage	n × 150	mV
01	70-cm transmitter average power	(253 − n)² / 2000	W
02	70-cm receiver temperature	(n − 127)/1.82	°C
03	(Reserved) 04 BCR output and main battery voltage	(n − 10) × 75	mV
05	(Special Purpose)	xxxxxxxxxxxx	
06	2-m transmitter power amplifier temperature	(n − 127)/1.82	°C
07	+14-V rail current to transponder	(n − 15) × 20.64	mA
08	+10-V regulator voltage	(n − 12) × 50	mV
09	He tank high pressure	(TBD)	Bar
0A	IHU temperature	(n − 127)/1.82	°C
0B	+14-V rail current to magnetorquers		mA
0C	BCR oscillator #1 status	0 = Off; N < 10 = On	—
0D	He tank low-side pressure-control voltage	(TBD)	—
0E	BCR temperature	(n − 127)/1.82	°C
0F	+10-V regulator current	(n − 15) × 4.128	mA
10	BCR oscillator #2 status	0 = Off, N < 10 = On	—
11	2-m-transmitter average power output	(200 − n)² / 2000	W
12	He tank temperature	(n − 127)/1.82	°C
13	SEU temperature	(n − 127)/1.82	°C
14	Battery charge current	(n − 15) × 10.32	mA
15	Top (+Z) photocell sun sensor	(See note below)	—
16	Motor-valve relay	(TBD)	—
17	Auxiliary battery #1 temperature	(n − 127)/1.82	°C
18	Active BCR output current	(n − 15) × 20.64	mA
19	Bottom (−Z) photocell sensor	(See note below)	—
1A	S-Band-transmitter power output		mW
1B	Active BCR input current on 28-V line	(n − 15) × 10.32	mA
1C	Spin rate {if n≠139, r = 508/(n − 116)-2		r/min
	or {if n < = 139 r = (139 − n) × 0.8 + 20		r/min
1D	24-cm receiver AGC {if n×100 AGC=0 or {if n < = 100 AGC = (n − 100)² / 189		dB
1E	Main battery temperature	(n − 127)/1.82	°C
1F	Special purpose	xxxxxxxxxx	—
20	24-cm receiver temperature	(n − 127)/1.82	°C
21	Solar panel #6 current	(n − 15) × 4.128	mA
22	He tank temperature	(n − 127)/1.82	°C
23	Solar panel #1 temperature	(n − 127)/1.82	°C
24	Solar panel #5 current	(n − 15) × 4.128	mA
25	70-cm receiver AGC	(n − 83)² / 1000	dB
26	70-cm transmitter PA temperature	(n − 127)/1.82	°C
27	Solar panel #3 temperature	(n − 127)/1.82	°C
28	Solar panel #4 current	(n − 15) × 4.128	mA
29	Special purpose	xxxxxxxxxxxx	—
2A	Solar panel #5 temperature	(n − 127)/1.82	°C
2B	Solar panel #3 current	(n − 15) × 4.128	mA
2C	+14-V regulator voltage	(n − 10) × 61.5	mV
2D	RUDAK temperature	(n − 127)/1.82	°C
2E	Solar panel #1 current	(n − 15) × 4.128	mA
2F	Solar panel #2 current	(n − 15) × 4.128	mA
30	Mode-B transponder +9-V supply voltage	(n − 10) × 50	mV
31	Wall temperature in arm #2	(n − 127)/1.82	°C
32	Bottom (−Z) skin temperature of arm #1	(n − 127)/1.82 C	°C
33	Special purpose #1 current	(n − 15) × 4.128	mA
34	Solar panel #1 current	xxxxxxxxxxxx	—
35	Wall temperature in arm #1	(n − 127)/1.82	°C
36	N₂O₄ tank temperature	(n − 127)/1.82	°C
37	Reserved		
38	Auxiliary battery voltage	(n − 10) × 75	mV
39	Mode-S transponder temperature	(n − 127)/1.82	°C
3A	+Z platform temperature	(n − 127)/1.82	°C
3B	(SERI experiment)	—	—
3C	Mode-L transponder +9 V supply voltage	(n − 10) × 50	mV
3D	AZ-50 tank temperature		°C
3E	Nutation damper temperature		°C
3F	Reserved		

Notes on Table 3

The equation values are preliminary estimates only.

IHU = integrated housekeeping unit; the computer
BCR = battery charge regulator
He = Helium
SEU = sensor electronics unit
N₂O₄ = nitrogen tetroxide; the propellant oxidizer; AZ-50 is aerozine-50, the propellant fuel
TBD = to be defined

Notes regarding channels 14 and 18:
These two sensors detect sunlight on the top (+Z) surface and the bottom (−Z) surface of the spacecraft. They provide only a rough indication of sun position and are used to resolve ambiguity in the readings from the sun sensors. When in full sunlight, a sensor count of 65 results. A count of about 10 is background noise only, and the sun is present when the count exceeds 20. The main spacecraft solar cell arrays receive maximum illumination when the sun is perpendicular to (normal to) the spin axis (Z axis). When this condition exists, both the +Z and −Z detectors will yield background readings (10) only. A count of 20 or greater on either indicates misalignment. The higher the count, the greater the misalignment. The spacecraft with respect to the sun angle (beta angle)—Source: Jan King, W3GEY, Apr 15 86; ASR No. 93, Dec 31 84; 1. T. Ashley, ZL1AOX, Jan 27 88.

The Control Subsystem

The control subsystem is perhaps the most complex of all. It consists of a computer and its associated memory, software, command detectors, sensors and attitude-control devices. The heart of the control subsystem is the integrated housekeeping unit (IHU). The IHU consists of an RCA COSMAC CDP1802 microprocessor, its memory and various supporting elements. This CPU was chosen because a radiation-hardened silicon-on-sapphire version was available. Even though much more advanced CPUs are now available, the 1802's comparatively light task loading allows it to loaf along. The memory for the CPU consists of special Harris HS-6564RH radiation-hardened CMOS ICs donated by Harris' Custom Integrated Circuit Division, Melbourne, Florida. The total memory carried is 48 kbytes, of which the error-corrected available net memory is 32 kbytes. The software is written in IPS, a FORTH-like multi-tasking language developed by Karl Meinzer, DJ4ZC, and his colleagues about 10 years ago. The IHU controls the satellite. Computer elements of the power subsystem provide closely regulated power to the other subsystems. Thus, the battery ensures that the satellite is powered even when in Earth's shadow.

The battery charge regulator (BCR) ensures the two nickel-cadmium batteries on board are optimally charged. The primary battery has a capacity of 10 Ah; the auxiliary battery is rated at 6 Ah. Together, the elements of the power subsystem provide closely regulated power to the other subsystems. Thus, the battery ensures that the satellite is powered even when in Earth's shadow.

Radiation damage incurred in orbit after the first three years is expected to reduce the total array power to about 35 watts.

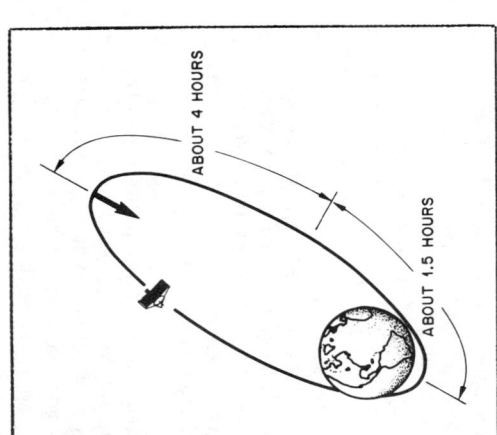

Phase 3C circles the earth in a semi-synchronous orbit called Molniya. The Molniya orbit is elliptical and has an orbital period of about 11 hours.

mands are uploaded from the ground to the IHU. The IHU then controls the satellite for weeks or months without intervention from ground controllers. The IHU turns the transponders on and off according to a prescribed schedule, carefully regulates the battery charge in conjunction with the BCR, and issues a continuous stream of telemetry via the communications subsystem to show sensor and status indicator readings on their own.

The IHU controls the attitude of the satellite to ensure it is aimed precisely at earth, so its antennas are properly illuminating the target area and so the solar panels are aligned toward the sun. To control the satellite's attitude, the IHU compares readings from its earth sensors and sun sensor with mathematical models contained in its memory. When it detects a deviation in attitude from the proper one, it applies carefully timed pulses to on-board electromagnets. The field from these magnets interacts with the earth's geomagnetic field, creating a torque that tends to reorient the spacecraft. When the spacecraft's attitude has been restored to the desired value, the magnetorquing ceases.

The IHU also maintains the proper spin rate for Phase 3C. Again using the magnetorquers, the IHU spins the satellite like the armature of a motor. The spinning stabilizes the satellite, just as spinning stabilizes a top. Spinning also ensures the satellite stays thermally manageable. Like the spit of a rotisserie, the satellite spins in the brilliant glare of the sun, absorbing heat when facing the sun, and radiating it to the blackness of space when facing elsewhere. Phase 3C will be spun at about 33 revolutions per minute for most of its operational life.

The IHU also schedules and transmits pre-stored bulletins on the various beacon frequencies. The bulletins usually consist of user-related information, such as the satellite operating schedules, orbital data, special advisories and so forth. The bulletins are sent in various modulation formats: (1) CW, (2) 45-baud FSK RTTY and (3) 400 bit/s PSK ASCII bulletins.

Temperature and pressure sensors, and various electrical sensors, feed a multiplexer that then feeds the IHU, which produces the 64-channel telemetry format. The telemetry suite (See Table 3) is comprehensive, allowing close monitoring of virtually all important aspects of satellite operation by command stations and suitably equipped ground observers. The earth sensors and sun sensor feed the sensor electronics unit (SEU) that the IHU uses to determine spacecraft attitude.

If it were not for the IHU, the satellite would be little more than a big hunk of dumb iron. With the IHU, the satellite is one of the most advanced Amateur Radio devices ever built. The fact that the satellite is totally under IHU software control ensures unprecedented satellite flexibility. The IHU can run the satellite for weeks without intervention from the ground command team.

The Propulsion Subsystem

The Ariane-4 rocket that carries Phase 3C to orbit is the largest launcher ever built by the European Space Agency. It will carry two large commercial geosynchronous satellites in addition to Phase 3C. But the Ariane-4 does not place any of the three payload satellites in their final orbits. Rather, each is deposited in its own geosynchronous transfer orbit (GTO—see Table 4). A GTO is the point of departure between the launcher and the payloads. The launcher orients itself, spins up the satellites and sets them off on their own. The GTO is a highly eccentric ellipse with a perigee around 220 km (137 miles) and an apogee of about 36,000 km (22,000 miles). Each satellite must get to a more usable orbit using its own rocket system called a *kick motor* (so named because it figuratively "kicks" the satellite from the GTO to a higher orbit).

For Phase 3C, the objective is to get from the GTO to a more stable orbit within a few days. There is some urgency in raising the perigee quickly, since satellites with 220-km perigees tend to fall from the sky regularly and, on occasion, in a great flurry of debris—literally as man-made meteorites. Since AMSAT folks fancy themselves builders of spacecraft and not ersatz meteorite makers, the designers have resolved to execute the first in-orbit burn as soon as the orbit parameters can be accurately determined. Using a special ranging technique and sophisticated mathematical models on powerful computers, AMSAT engineers will be able to develop GTO orbital data for Phase 3C within a few days of launch.

The perigee kick motor (PKM) on Phase 3C is a small rocket engine that uses two fluids stored in separate portions of a fuel tank. Under pressure from a helium system, the propellant and oxidizer combine in the combustion chamber and spontaneously ignite in a controlled explosion to produce a thrust of 400 Newtons (about 95 lbs). A 400-N thrust would not move your Buick up the highway that smartly, but on the business end of a 140-kg (309-lb) satellite, the result is moderately impressive.

In a series of kick-motor burns spread over a few weeks, the perigee is raised to about 1500 km and the inclination of the plane of the orbit is raised from the 10° GTO to about 57°. The total velocity change imparted by the little Messerschmitt-Boelkow-Blohm (MBB) engine amounts to an impressive 1480 m/s (4854 ft/s). To put that number in perspective, that's the equivalent of going from rest to 3310 mi/h (Mach 5). (Phase 3C could outrun an SR-71 Blackbird superspy jet!) Fortunately, the acceleration imposed on Phase 3C is not overwhelming: At just 3Gs, it amounts only

Table 4
Orbital Characteristics (at separation from launcher)

Geosynchronous transfer orbit (GTO)
Perigee altitude:	222.504 km
Apogee altitude:	36076.636 km
Inclination:	9.997 degrees
Argument of perigee:	178.148 degrees
Ascending node longitude:	−135.541 degrees/lift-off
True anomaly:	127.554 degrees
Epoch:	Instant of Separation (L +4797.1 s)
Spin rate at deployment:	29.47 degrees per second
Separation velocity:	0.59 meters per second

Objective Phase 3C orbit (final orbit after two or three burns)
Apogee:	36,000 km
Perigee:	1,500 km
Inclination:	approx 57 degrees
Argument of perigee:	178 degrees (determined by launcher)
Anomalistic period:	approx 662.4 minutes
Longitude increment:	approx 184.5 degrees East per orbit

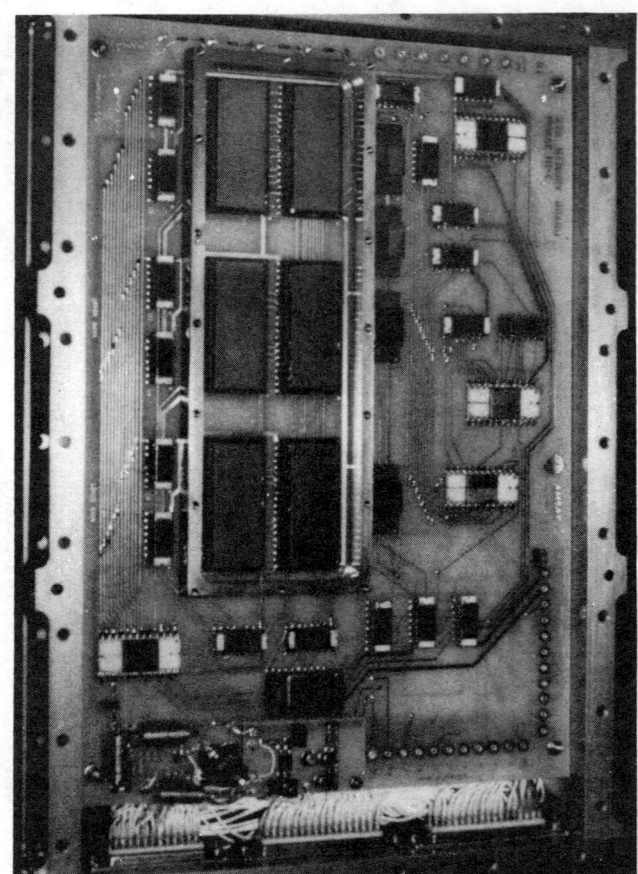

The Phase 3C integrated housekeeping unit, a sophisticated computer that controls the satellite, now contains radiation-hardened memory donated by Harris. The IHU was designed by Steve Robinson, W2FPY, and built by Gordon Hardman, KE3D. (photo by Dick Daniels, W4PUJ)

to a modest imposition for which the designers have taken adequate provision. The rocket engine can be restarted several times to refine the orbit until the desired result is obtained. All burns, however, should be completed within a month of launch.

What's To Do On Phase 3C

Since Phase 3C is primarily a spaceborne communications system, logic suggests communications is the main activity. Logic holds here. The Mode-B transponder provides a useful training ground for those learning the ropes. Its 150 kHz of available bandwidth is adequate for about 50 simultaneous QSOs under optimum conditions. Since most satellite contacts involve more than two participants, Mode B can actively serve more than 100 users at a time.

The equipment required to use Mode B is reasonably modest, and many individuals can get the feel of satellite operations on the new bird with borrowed equipment or that already on hand. Rag chews, DXing, nets and general socializing are popular. Round table discussions are a favorite of mine. Getting a half-dozen chums from across the globe together for an easygoing, relaxed rag chew is very enjoyable. Annoying QRM is virtually absent and QSOs are clearly audible. Conditions are comparable to those on the 20-meter band at its peak, as if you had a large swatch of it to yourself. With coverage times measured in hours, there's no need to rush QSOs as was necessary with low earth-orbiting satellites. And since Phase 3C moves slowly across the sky when near apogee, antennas needn't be moved much,

if at all. Table 5A details ground station requirements for Mode B.

The Mode-JL Transponder

Although Mode B may become the mode

of choice for satellite beginners, Mode JL seems destined to become the most popular and productive. Using combined uplinks on 24 cm and 2 meters, Mode JL has the right combination of attributes to delight the satellite veteran and newcomer alike. It combines a passband broad enough (290 kHz) for perhaps 90 simultaneous SSB QSOs with other attractions bound to make it a pleasure to use. The uplink power requirements on 24 cm are quite modest, as Table 5B shows. Since antennas can be small and power requirements are modest, I suspect Mode JL will become popular with amateurs wanting to operate from remote locations (campsites and mountain-tops). Whereas the 24-cm Mode-JL uplink offers the prospect of portable antennas, the 2-meter Mode-JL uplink is designed for lesser-developed nations where 24-cm equipment is rare or nonexistent.

The Mode-S Transponder

Mode S will be a special delight to use if all goes according to plan. The Mode-S experiment is designed for a single FM channel. The actual use of this channel hasn't been determined, but some ideas have been suggested. Besides the obvious QSO activities between two or more individuals, the Mode-S transponder has been suggested for occasional use as a bulletin channel. The high-quality audio that can likely be received at well-equipped Mode-S stations could be easily patched into local terrestrial repeaters to furnish a

Table 5A
Minimum Mode B Station Requirements

Uplink
Frequency: 435.420-435.570 MHz
EIRP: 21.5 dBW (141 W) for 20 dB peak and 10 dB average SNR on downlink
Polarization: Right Hand Circular (RHC)
Suitable uplink components 10 watts to 12 dBic gain antenna

Downlink
Frequency: 145.975-145.825 MHz
Polarization: RHC
Minimum recommended antenna gain: 10 dBic
Maximum receive system effective noise temperature: 625 K (NF = 5.0 dB)
Minimum figure of merit: −18 dB/K

Table 5B
Minimum Mode JL Station Requirements

Uplink
Frequency: Mode L: 1269.620-1269.330 MHz
              Mode J: 144.425-144.475 MHz
EIRP: Mode L: 25 dBW (316 W) for 20 dB peak and 10 dB average SNR on downlink
      Mode J: 25 dBW (316 W) for 20 dB peak and 10 dB average SNR on downlink
Polarization: RHC
Suitable uplink components: Mode L: 10 watts to 15 dBic gain antenna
                            Mode J: 20 watts to 12 dBic gain antenna

Downlink
Frequency: 435.715-436.005 MHz
Polarization: RHC
Minimum recommended antenna gain: 13 dBic
Maximum receive system effective noise temperature: 290 K (NF = 3.0 dB)
Minimum figure of merit: −12 dB/K

propagation zones appear and disappear. Not so on the OSCAR satellites! On the transponder, everyone listening on a given frequency hears everyone else using that frequency equally well (more or less, assuming comparable equipment). There is no frequency sharing afforded by geography (space diversity). The satellite illuminates its footprint evenly. Consequently, the potentially disruptive effects of high-intensity contests are magnified within the bounds of the transponder passband.

Second, the competitive spirit of contesting on the HF bands leads some participants to use copious amounts of power. Increasing power on HF increases the coverage zone. The higher the power, the larger the zone covered, and the less are the possibilities of frequency sharing. (Some high-power Navy VLF signals are heard the world over, even by submerged submarines!) Barring untoward splattering, raising your power on 20 meters does not, however, affect a QSO 100 kHz away.

Unfortunately, the same cannot be said for satellite operation. A finite amount of power is available to the transponder. Since the transponders are linear, the manner in which the available downlink power is divided among passband users is in direct proportion to the strength of their uplink signals. If one very strong uplink signal appears in the passband, it uses an inordinate amount of the available downlink power and, consequently, reduces the power available to the other users. In contrast to HF operation, therefore, an overpowered station 100 kHz down the band *can* unduly affect your QSO. "Power hogging" is an abomination to the conscientious satellite operator and, since contests seem to promote power escalation scenarios, contests have been discouraged on all OSCARs.

But there is more to on-air competition than contests. DXing on satellites is a form of long-term competition that is reasonable insofar as it, too, stays under control. DXing and seeking awards for grid squares, countries or whatever is normally a lower-keyed form of competition and is encouraged and promoted with various satellite awards. Furthermore, a new type of competitive activity was begun on AO-10 and will soon be resumed and enhanced on Phase 3C. This new competitive activity is called *Techno-Sport*. It's a variation on the highly popular radio sport concepts of the USSR and other countries. In AMSAT's Techno-Sport activities, the idea is to harness the competitive drives of individuals in a way that promotes learning about technical matters and development of useful technical skills.

For example, AMSAT's *ZRO Test*[5] got started in 1985 on AO-10. The challenge of the ZRO Test is to develop and operate a superior Mode-B or Mode-L station that can "hear" exceedingly well. Through carefully calibrated and controlled conditions, an uplink signal is sent to the satel-

Table 5C
Minimum Mode S Station Requirements

Uplink
Frequency: 435.601-435.637 MHz
EIRP: Approx 27 dBW (501 W) under average Mode B AGC conditions
Polarization: RHC
Suitable uplink components: 25 watts to 13 dBic antenna

Downlink
Frequency: 2400.711-2400.747 MHz + beacon @ 2400.325 MHz
Polarization: RHC
Minimum recommended antenna gain: 28 dBic
Typical antenna: 1.4 m dish (assuming 50% efficiency)
Maximum receive system effective noise temperature: 290 K (NF = 3.0 dB)
Minimum figure of merit: +3 dB/K

fine bulletin distribution system. This might be viewed as a prototype for operational planning for AMSAT's Phase 4 geosynchronous satellites planned for the early 1990s. Up to four SSB stations can use the Mode-S transponder, too. Ground station requirements for Mode S are detailed in Table 5C.

The Rudak Project

RUDAK is essentially a digipeater designed and built by AMSAT DL engineers in Munich. The architects of RUDAK envision it providing an interconnect between local area networks in addition to point-to-point contacts. An independent receiver in the Mode-L transponder is provided for the RUDAK uplink on 1269.710 MHz. The demodulator converts the 2400-bit/s biphase PSK signal into a clean digital signal for the RUDAK processor. Thanks to the sweep circuit in the demodulator, the uplink signals only have to be within ±7.5 kHz of the center frequency.

On the downlink side, the output data modulates the RUDAK beacon transmitter in the Mode-L transponder on 435.677 MHz using BPSK at a data rate of 400 bit/s, the same as for the general beacon of Phase 3C. The rate can be increased to 1200 bit/s using NRZI (nonreturn to zero) modulation for experimental reasons. Details on interfacing to RUDAK are beyond the scope of this article and have been published elsewhere.[4]

Some basic RUDAK station requirements are provided in Table 5D.

QSOs, Competition and Techno-Sport

Satellite activities other than one-on-one QSOs, round tables or nets are organized by AMSAT and other groups. Emergency communications exercises have been carried out on earlier OSCARs and will continue on Phase 3C. On the other hand, linking terrestrial repeaters to the hemispheric coverage of Phase 3C should provide some excitement for many whose only radio is a 2-meter or 70-cm FM handheld transceiver. Imagine talking to someone in Australia from the comfort of your easy chair while holding your hand-held transceiver! Gateways provide a means of consolidating resources so that many can enjoy the thrill of satellite communications without making a large investment in equipment. When users find out how much fun it can be, they may opt to establish their own satellite stations. Gateways are, in essence, a means of tasting the wine before buying the bottle.

Competition on OSCARs has always been approached with caution. There are good reasons for this. First, transponder operation is different from the HF bands. On HF, there is a degree of frequency sharing. If you turn your HF beam away from an interfering station, you can reduce the QRM and continue your QSO. Various

Table 5D
Minimum RUDAK Station Requirements

Uplink
Frequency: 1269.710 MHz
EIRP: 26 dBW (400 W)
Typical suitable uplink: 8 watts to 17 dBic antenna
Polarization: RHC

Downlink
Frequency: 435.677 MHz
Typical receive antenna gain: 10 dBic for 12 dB E_b/N_o ratio†
Polarization: RHC

†Energy per bit to noise ratio

AMSAT PHASE III C

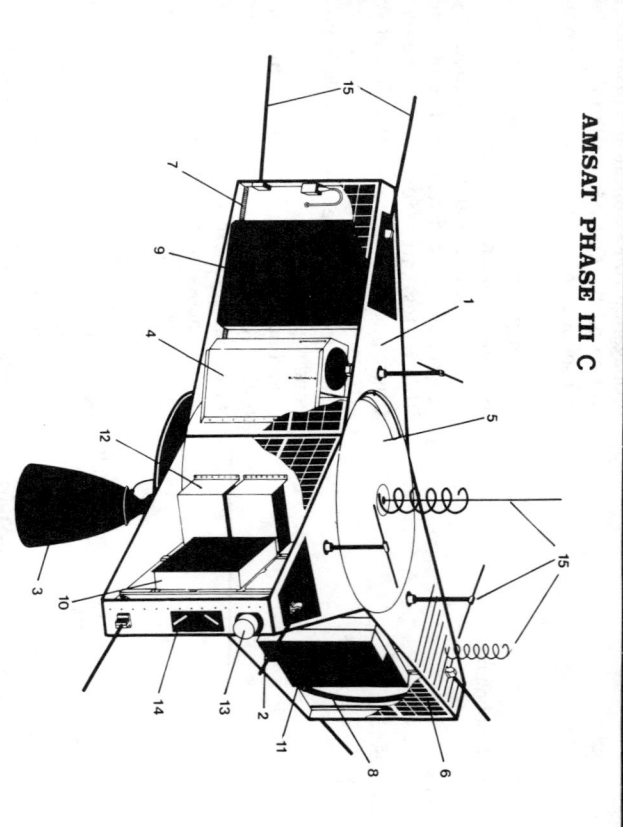

The AMSAT Phase 3C satellite. (graphic by AMSAT-DL)

1—Aluminum space frame
2—S-band transponder
3—Kick motor
4—Helium tank container
5—Fuel and oxidizer tank
6—Solar panel
7—Magnetorquer coil
8—Nutation dampener
9—Integrated housekeeping unit
10—Battery charge regulator
11—Modulator
12—Auxiliary battery
13—Earth sensor
14—Sun sensor
15—Antennas

Thus, there's plenty to do on Phase 3C. With all the activity that's on tap, contests are scarcely missed.

Operating Tips

Operating via any satellite is a snap once the basics are mastered. True, there are a few new things to learn to become proficient on the OSCAR satellites, but learning new techniques is part of the enjoyment of satellite operation. To successfully operate via the OSCAR satellites, you should have knowledge of five basic areas:

1) Where the satellite is
2) What mode it's in
3) What frequency to transmit on
4) What frequency to receive on
5) Basic operating practice

Determining "where the satellite is" is called *tracking*. Since high-gain antennas are required, you need to aim them at the satellite properly. Tracking tells you where to aim. Manual tracking devices are convenient, inexpensive and sufficiently accurate for satisfactory results. Tracking with the aid of a computer is by far the easiest—just enter the data on the satellite's orbit.[8]

Item 2, mode of operation, is a matter of scheduling. The operating schedule of Phase 3C will be announced after launch. You can learn the operating schedule by listening to W1AW bulletins, by obtaining official AMSAT net bulletins via on-air nets, from packet-radio bulletin boards, or by land-line bulletin boards.[9]

Items 3 and 4, transmit and receive frequencies, are addressed in Table 1. More specific operating information comes in the form of band plans. The preliminary idea is to divide the satellite passbands into thirds, as was done for AO-10. The lowest third is for digital modes such as CW and packet, the upper third is for voice modes such as SSB, and the center third is for mixed modes.[10]

Item 5, basic operating practice, reduces to some very straightforward guidance. For example, always wear headphones when operating SSB. Why? Since you're operating full duplex, a feedback path will exist between your receiver's speaker and the transmitter microphone. The feedback path will round through the satellite back to you again. Howling feedback soon alerts you to the need to don the "cans." Next, never swish the transmitter across the passband to locate your downlink frequency. Estimate your downlink frequency by using Table 1. Then, send a short series of CW dots or briefly whistle into you microphone while listening carefully for your downlink signal and tuning your receiver slightly. To correct for Doppler shift on Mode B and Mode JL, adjust your transmitter frequency only. On Mode S, adjust your receiver frequency only. General operating guidelines appear in AMSAT's *Phase 3 Operating Handbook*. This Handbook is now being revised for Phase 3C and is due for completion soon. Experience is a good

lite. All listeners receive equal downlinks (within a fraction of a decibel) because of the position of the satellite. After a practice run, a series of CW numbers is sent at the baseline signal strength level. The baseline downlink level is set to equal the beacon. The power output of the beacon remains constant in both the short and long term. This ensures level correlation from month to month between test sessions and should reduce calibration errors because of variations in satellite attitude and ionospheric absorption.

ZRO Test participants carefully copy the numbers sent in slow CW. The uplink power is then reduced by half (3 dB). More CW numbers are sent and copied. Then power is reduced by another 3 dB and more numbers are sent and copied. The process continues in 3-dB increments until the CW on the downlink is so weak that virtually no one copies it (at 24 dB below the baseline level, the uplink is less than 1 watt!). Certificates are awarded to those who copy the baseline numbers. Endorsement stickers are awarded each time a competitor demonstrates that he or she has improved his or her receive sensitivity by 3 dB. Only one competitor (Jeffrey Bishop, W7ID) has ever garnered the long-sought Z-8 award—24 dB below the baseline.

New Techno-Sports are being planned for Phase 3C. There's *SatFox*. This is a derivative of the terrestrial radiolocation sport, Fox and Hound. In the satellite version, you and your computer try to figure out where the hidden transmitter is. Instead of traipsing through the swamps with your DF loops, however, you play hound from the comfort of your shack.

SatFox will use the movement of the satellite itself to obtain "fixes" on the hidden fox who could be 80 or 8000 miles away. If you're a clever hound, you should be able to narrow the fox's QTH down to a county-sized area. You send in your estimate by mail. If you do well, you earn a handsome award or an endorsement to a prior award.

The SatFox location technique will use Doppler-shift measurements to isolate a hidden RF source, rather than signal strength (used in some terrestrial Fox and Hound competitions). Doppler shift measurements are the basis of the search-and-rescue satellite system now being operated by more than a dozen nations using the SARSAT/COSPAS satellites of the US and USSR.[6] Project SatFox on Phase 3C will work in an analogous way. The fox will transmit a beacon, but since the motion of Phase 3C is low compared to SARSAT/COSPAS satellites, extra care will be required to make precise position determinations. That's where skill comes in. And that's what Techno-Sport is all about: combining the fun of competing with the satisfaction of learning and the occasional reward of prevailing and being recognized for superior performance.

ship in the amateur space program.

The AMSAT-DL team members work on Phase 3C during integration in Golden, Colorado; (Left to right) Konrad Mueller; Hanspeter Kuhlen, DK1YQ; Karl Meinzer, DJ4ZC; Werner Haas, DJ5KQ. (AMSAT-DL photo)

teacher, and many helpful colleagues can be found on OSCARs, since they feel proud of having joined the space-age hobby. They are often glad to share their findings with you!

I hope you have gathered that becoming proficient on OSCAR, even Phase 3C, is no big deal. Sure, there is some esoteric information involved, but Amateur Radio in general probably appeared fairly mysterious to you until you got thoroughly involved. Identifying and using reliable information sources is the key.[11]

In this article, we've learned what Phase 3C is and what it can do. We've recognized that getting to know how to use it and have a great deal of fun in the process is simply a matter of getting interested and having access to the right information. Products and activities of the space age are all around us in our everyday lives. Getting your signal onboard OSCAR and being a part of the Amateur Radio space program is as natural for hams as can be. Phase 3C could be your entry into this vast, rewarding new realm of radio communications. See you there!

Acknowledgments

The propulsion section is based on Richard L. Daniels, W4PUJ, "The Propulsion Systems of the Phase-III Series Satellites," *Proceedings of the 1st Utah State University Conference on Small Satellites*, Oct 7, 1987. (This work is reprinted in the *AMSAT-NA Technical Journal*, Vol 1, No. 2, Winter 1987-88, published by AMSAT-NA).

The IHU section is based on Gordon Hardman, KE3D, "The Integrated Housekeeping Unit—A Method of Telemetry, Command and Control for Small Spacecraft," *Proceedings of the 1st Utah State University Conference on Small Satellites*, Oct 7, 1987. (This article is also reprinted in the *AMSAT-NA Technical Journal*, Vol 1, No. 2, Winter 1987-88, published by AMSAT-NA).

I am especially grateful to AMSAT's Vice President of Engineering, Jan King, W3GEY, for help in preparing this article and for his leader-

Notes

[1] OSCAR stands for Orbiting Satellite Carrying Amateur Radio. OSCAR 1 was launched in 1961. President Gerald Ford's voice (tape delayed for the next orbital pass) was heard via AMSAT-OSCAR 7 on July 1, 1976. King Hussein of Jordan, JY1, has operated via AMSAT-OSCAR 10. Used as a generic term, OSCAR includes all the individual OSCARs and the Russian RS (Radio Sputnik) series as well. In a slightly more abstract sense, OSCAR connotes the notion that there's a place for direct public access to space and space-based activities, that is, the Amateur Radio space program.

[2] The Russian RS10/11 and the Japanese FO-12 satellites carry on the Phase 2 satellite tradition.

[3] Circular polarization is used extensively in space communication to offset the effects of polarization rotation in the geomagnetic field and to reduce the effects of the spinning spacecraft. Circular polarization is covered in the ARRL publication *The Satellite Experimenter's Handbook* and is recommended reading.

[4] RUDAK description quoted from an article by Peter Guelzow, DB2OS, appearing in *Amateur Satellite Report* No. 126/127, June 24, 1986. Also, see the QST Amateur Satellite Communications column for July and August 1988.

[5] The ZRO Test is the ZRO Memorial Satellite Station Engineering Award. It is a memorial to Kaz Deskur, K2ZRO, who, until his death in April 1984, was a strong AMSAT technical resource person.

[6] SARSAT/COSPAS is the same system that was used to track the Russian and Canadian SKITREKers on their journey across the North Pole beginning in March. A small emergency locator transmitter (ELT) signal was picked up by the SARSAT/COSPAS satellites. On the ground, the Doppler shift of the resulting downlink was analyzed and the trekers' position closely determined by plotting various curves based on the Doppler shift measurements.

[7] I've covered many of these topics in my Amateur Satellite Communications column appearing in QST over the last few years.

[8] Orbital data originates with NASA. Two distributors of orbital elements are the ARRL and AMSAT. Tracking software is available from the AMSAT Software Exchange, PO Box 27, Washington, DC 20044. Send an SASE for a list of available programs.

[9] Operating schedules in general were discussed in the Amateur Satellite Communications column in April 1988 *QST*.

[10] An international band-planning committee is being considered to help develop a Phase 3C band plan.

[11] AMSAT members receive regular newsletters and access to helpful publications, such as the *AMSAT-NA Technical Journal*.

Adventures In Satellite DXing

Part 1: Crowded HF bands and marginal propagation got you down? Use the high-flying DX machine to give your spirits a lift.

By Dick Jansson, WD4FAB
1130 Willowbrook Trail
Maitland, FL 32751

Is there such a thing as an impossible dream? Imagine relaxed ragchewing with exotic DX stations on SSB without the congestion found on 20 or 75 meters, all with only 15 watts of RF power. Not possible? Sure it is—just try your hand at OSCAR 10.

Oh sure, here we go again, another lecture on amateur satellites! Hold on there—don't turn the page yet! We are not going to talk strictly about satellites, but about how you can enjoy the fun of operating a super DX machine at a casual pace that won't push your stamina or heart! For example, one recent morning I talked with hams in Texas, Colorado, Antarctica, England, Massachusetts, Grenada and South Africa. All of this occurred when some operators were bemoaning that the satellite wasn't too active. Interested? This is the first of a four-part series of articles that will show you what satellite operating is all about. This installment will explain what OSCAR 10 is and what it can do for you. In the coming months, we'll take a look at the equipment and antennas for satellite work and show how to assemble them into a station that really plays. Then we'll examine some of the finer points of communicating through OSCAR 10.

Not for Beginners Only

If you think that this discussion is aimed at beginners, you are very right. All of us, even old-timers, are "beginners" at the satellite game at one time or another. Part of the beauty of satellite operation is that there's always something new to learn. Just this year I've picked up some new tricks even though I've been operating amateur satellites regularly since 1977. This form of communication is very different from the HF bands.

As we proceed, I'll try to avoid some of the involved jargon of satellite talk. Where appropriate I'll explain what the terms mean so you'll be able to hold an intelligent conversation with others interested in OSCAR operation. I do highly recommend that every satellite user obtain and read a copy of *The Satellite Experimenter's Handbook* by Martin Davidoff, K2UBC.[1] This fine publication is packed with useful information that ranges far beyond the scope of this *QST* series.

One special term that needs to be explained up front is AMSAT.[2] This is the name given to the Radio Amateur Satellite Corporation, a nonprofit, scientific corporation founded for the creation and operation of satellites for Amateur Radio communications. Over the years, AMSAT has continued to make reliable communications satellites available to the amateur community. AMSAT's paid staff is very small. Hundreds of volunteers support the organization's efforts through donations of time, expertise and money, making possible the superb communications available from OSCAR 10 and other spacecraft. Every amateur-satellite user should join AMSAT to show their support for the organization and to help with the construction of the satellites we use. AMSAT members receive *Satellite Journal*, a publication filled with useful information for everyone. New members also receive a copy of *A Beginner's Guide to OSCAR 10*, which is full of information helpful to the newcomers and old hands alike.[3]

What Is OSCAR 10?

At the risk of boring the more knowledgeable readers, I'll give a short history of OSCAR 10. Some of you may want to skip to the next section.

OSCAR 10 (also called AO-10 for AMSAT-OSCAR 10) is the tenth in a series of amateur satellites. The name OSCAR, for *Orbiting Satellite Carrying Amateur Radio* pretty much tells the story. It is a communications satellite designed and built by radio amateurs for the sole purpose of supporting Amateur Radio communications and experimentation. It's very much like the satellites that allow you instant telephone access to relatives overseas and allow you to watch televised events occurring on the other side of the world as they happen. OSCAR 10 receives transmissions from earth stations and relays them back, allowing hams to communicate over great distances without worrying about the whims of ionospheric propagation.

Fig 1 shows OSCAR 10 pretty much as it looks today. Weighing about 200 pounds, this satellite was launched by a European Space Agency rocket in June 1983. Although it is an "amateur" satellite, there is nothing amateur about OSCAR 10's design and construction. It is built to the same high standards as other communications satellites and was subjected to strenuous testing before launch. Historically, amateur satellites have provided reliable service well beyond their design lifespans.

About Orbits

Some of you may have operated through OSCARs 6, 7 and 8. They were in circular orbits less than 1200 miles above the earth.

Fig 1—OSCAR 10 is small enough that it can be carried by two or three people. Much of the outer surface is covered with solar panels that will power the satellite for years.

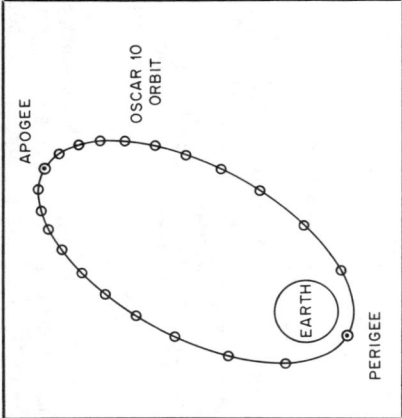

Fig 2—OSCAR 10 was designed for an elliptical orbit to allow it to be in view of much of the Northern Hemisphere for hours at a time. The circles on the ellipse represent the approximate position of OSCAR 10 during each half hour of its orbit. Note that the satellite "slows down" as it nears apogee. It passes through perigee (where it is out of sight of stations in the Northern Hemisphere) rather quickly.

Glossary

AMSAT—the Radio Amateur Satellite Corporation, a nonprofit organization located in Washington, DC has overseen the OSCAR program since the launch of OSCAR 5.
AO-10—AMSAT-OSCAR 10, the 10th amateur satellite in the OSCAR series.
apogee—the point in a satellite's orbit at which it is farthest from the earth.
downlink—the frequency on which signals are transmitted from the satellite to earth.
elliptical orbit—those orbits in which the satellite path traces an ellipse with the earth at one focus.
Mode A—transponders with a 2-meter uplink and a 10-meter downlink.
Mode B—transponders with a 70-cm uplink and a 2-meter downlink.
Mode L—transponders with a 24-cm uplink and a 70-cm downlink.
OSCAR—Orbiting Satellite Carrying Amateur Radio.
pass—that segment of a satellite orbit that brings it in range of your station.
passband—the range of frequencies handled by a transponder.
perigee—that point in a satellite's orbit at which it is closest to the earth.
Phase II—the name given to low-altitude orbit (less than 1200 miles) OSCARs. Equipped with solar cells, these satellites have lasted for up to five years.
Phase III—the name given to extended-range, high-altitude OSCARs in elliptical orbit.
transponder—a device that receives radio signals in one segment of the spectrum, amplifies them, translates (shifts) their frequency to another segment of the spectrum and retransmits them.
uplink—the frequency on which signals are transmitted to the satellite from earth.

SSB. The other big difference is that OSCAR 10's transponders cover not just a single frequency, but a whole range of frequencies. This range of frequencies is called a passband; the passband may cover 100 kHz or more. Amateur satellites retransmit every signal heard in the receiver passband, so many stations can use the transponder simultaneously.

The frequency on which you transmit to the satellite is called the uplink, while the frequency on which the satellite retransmits your signal to earth is called the downlink. Uplink and downlink frequencies are on different bands. Various band combinations are called modes. For example, Phase II satellites carried a Mode-A transponder that used 2 meters for the uplink and 10

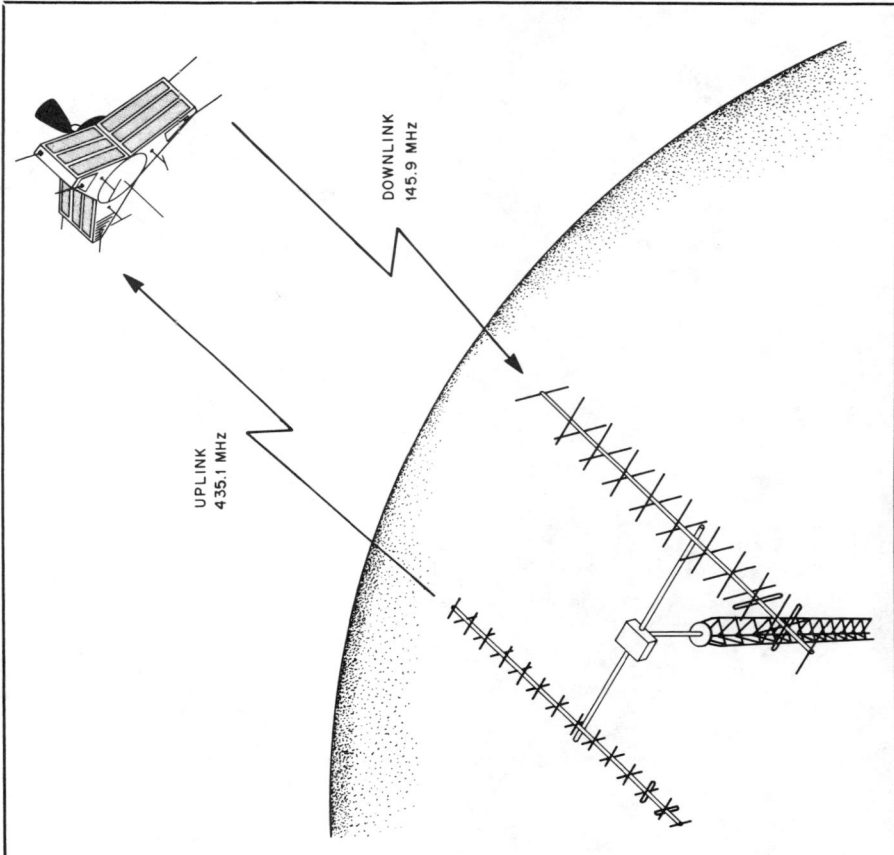

Fig 3—This series of articles will concentrate on Mode-B operation, whereby stations on earth transmit to the satellite on 435 MHz and listen on 145 MHz.

Known as Phase II satellites, these OSCARs provided communications over distances up to about 4500 miles. They were in range of a given point on earth for at most 20 minutes or so during each orbit. Phase II satellites moved quickly, so operators had to work hard to keep their antennas pointed at the "birds." One consequence of these 20-minute passes was that contacts tended to be short, contest-like exchanges of signal report, name and QTH.

Looking for something that would support communications for longer periods over longer distances, AMSAT designers developed a new generation of satellites, Phase III. As shown in Fig 2, these satellites have an elliptical orbit, rather than a circular one. At the apogee, the point in its orbit where it is farthest from earth, OSCAR 10 is about 22,000 miles away. At perigee, when it is closest, OSCAR 10 is about 2500 miles from earth.

In practice, this means that maximum communications distance via OSCAR 10 is more than 10,000 miles when the satellite is at apogee. It is accessible to nearly half the earth! Because the orbit is elliptical, OSCAR 10 is in view of stations in the northern hemisphere for up to eight hours at a time. Being able to work stations half a world away for up to eight hours at a time is my idea of relaxed DXing!

Links: Up and Down

OSCAR 10 carries two linear translators, or transponders. A transponder acts like a typical VHF FM repeater. The basic idea is the same: You transmit to it on one frequency, and it retransmits your transmission on another. There are some major differences, however. Unlike FM repeaters that are designed for nonlinear modes only, OSCAR 10's transponders are linear. They faithfully reproduce all modes, including

meters for the downlink.

OSCAR 10 carries a Mode-B transponder and a Mode-L transponder. The Mode-B uplink is at 435 MHz, while the downlink is at 145 MHz (see Fig 3). The passband is 152 kHz wide. Mode L uses a 1269-MHz uplink and a 436-MHz downlink. The passband is 800 kHz wide—more than twice as wide as the 20-meter ham band.

Most of the activity on OSCAR 10 is on Mode B, so I'll limit the scope of this article to that mode. Mode L presents a different set of equipment requirements. After you get some Mode-B experience, you may want to try your hand at Mode L. By that time, you'll know whom to contact for advice.

Some of the hard-core HF DXers will choke thinking that this VHF/UHF stuff is too complicated and want to return to 20 meters. Not so! The VHF and UHF parts actually make life much easier than you might think. There is a wealth of readily available commercial equipment for satellite work, as we'll see next month.

Notes
[1] M. Davidoff, *The Satellite Experimenter's Handbook* (Newington: ARRL, 1984). Available from your local radio store or from ARRL for $10 ($11 outside US). Add $2.50 ($3.50 UPS) per order for shipping and handling.
[2] AMSAT, PO Box 27, Washington, DC 20044. Dues are $24 per year.
[3] M. Crisler, *A Beginner's Guide to OSCAR 10* (Washington, DC: AMSAT, 1985).

Adventures in Satellite DXing

Part 2: In this installment, we'll take a look at some of the equipment you'll need to launch your OSCAR career.

By Dick Jansson, WD4FAB and Mark Wilson, AA2Z
1130 Willowbrook Trail   Senior Assistant Technical Editor, ARRL
Maitland, FL 32751

Last time we looked briefly at OSCAR 10's history and learned some basic terminology. Now it's time to get down to the specifics of the equipment you'll need to work stations through the bird. This month we will examine your options as you choose a receiver, transmitter and antenna system for OSCAR 10 operation.

The basic requirements for OSCAR 10 Mode-B operation are a sensitive 145-MHz receiver, a 435-MHz transmitter that can supply about 50 W of RF output and a high-gain antenna for each band. The antennas must be able to rotate in azimuth (side to side) and elevation (up and down). In earlier times, the satellite communicator had to work hard to assemble a usable station. These days, however, amateur equipment manufacturers provide some nifty boxes that make the job easier.

Full-Duplex Operation

Perhaps the biggest difference from terrestrial work is that satellite work requires full-duplex operation. This means that you transmit and receive simultaneously. You can hear your own downlink signal while you're transmitting, as well as that of the station you're working. Full duplex provides the opportunity for a fully interactive conversation, as if the other station is in the very same room! Gone are the endless monologues, and in come new communications methods.

The ability to hear your own signal on the downlink offers several advantages. You are assured that you and the station you want to work will be able to get on the same frequency. If you're responding to a CQ, you'll hear your signal and know that you're tuned to the right part of the passband. In addition, if you can hear yourself, others can copy you as well. There is no question about whether or not your signals are getting through. Moreover, you will know if you're running too much power and being a "satellite hog." It's possible to have too strong an uplink signal, as we'll see in a later installment.

Successful satellite operation demands that you can locate and hear your own signal from the spacecraft. You should choose equipment with this goal in mind. Equipping your station for full duplex operation is easier than you might think because the transmitter is on a different band than the receiver. Fig 1 shows several different satellite ground-station equipment configurations. Each of these options is discussed in the following pages.

Receivers

Receiving requirements for OSCAR 10 are stiff, but you can achieve pleasurable results with the right kind of equipment. Do not expect to find 60-dB-over-S9 signals on the downlink. OSCAR operation is a weak-signal situation where contacts can be made with signals that are 4 dB stronger than the noise. Conversational quality can be assured with signals that are 6-9 dB or greater out of the noise.

The old adage "You can't work 'em if you can't hear 'em" especially applies to satellite work. The first step you should take toward gearing up for OSCAR 10 is to assemble the best receiving setup you can. There's no point in getting transmitting capability until you can hear signals, and hear them well.

There are a number of options open to you if you are starting from scratch. You may wish to try your hand at building receiving equipment, or you may wish to purchase everything. All of the necessary components for the 145-MHz downlink are readily available from *QST* advertisers.

If you're active on 2 meters with a multimode transceiver, you already have the basic building block of your receiving setup. If you don't have any equipment but think you would like to try terrestrial SSB operation on 2 meters, you should consider purchasing one of these all-mode radios. The basic requirements are that the rig includes SSB and CW modes and that it covers the entire 2-meter band. A

Table 1
Suppliers of Equipment of Interest to Satellite Operators

Multimode VHF and UHF Transceivers and Specialty Equipment
ICOM America, Inc, 2380-116th Ave NE, Bellevue, WA 98004.
Ten-Tec Inc, Sevierville, TN 37862.
Trio-Kenwood Communications, 1111 West Walnut St, Compton, CA 90220.
Yaesu Electronics Corp, 6851 Walthall Way, Paramount, CA 90723.

Converters, Transverters and Preamplifiers
Advanced Receiver Research, Box 1242, Burlington, CT 06013.
Angle Linear, PO Box 35, Lomita, CA 90717 (preamps only).
Hamtronics, Inc, 65-E Moul Rd, Hilton, NY 14468.
Henry Radio, 2050 S Bundy Dr, Los Angeles, CA 90025.
The PX Shack, 52 Stonewyck Dr, Belle Mead, NJ 08502.
Radio Kit, Box 411, Greenville, NH 03048.
Spectrum International, PO Box 1084, Concord, MA 01742.
Transverters Unlimited, Box 6286, Station A, Toronto, ON M5W 1P3.
The VHF Shop, 16 S Mountaintop Blvd, Rt 309, Mountaintop, PA 18707.
TE Systems, PO Box 25845, Los Angeles, CA 90025.

70-cm Power Amplifiers
Alinco Electronics, PO Box 20009, Reno, NV 89515.
Communications Concepts, Inc, 2648 North Aragon Ave, Dayton, OH 45420.
Encomm, 1506 Capitol Ave, Plano, TX 75074.
Falcon Communications, PO Box 8979, Newport Beach, CA 92658.
Mirage Communications, PO Box 1000, Morgan Hill, CA 95037.

Antennas
Cushcraft Corp, 48 Perimeter Rd, Manchester, NH 03108.
KLM Electronics, Inc, PO Box 816, Morgan Hill, CA 95037.
Telex/Hy-Gain, 9600 Aldrich Ave South, Minneapolis, MN 55420.

Note: This is a partial list. The ARRL and *QST* do not endorse specific products.

Satellite Anthology 13

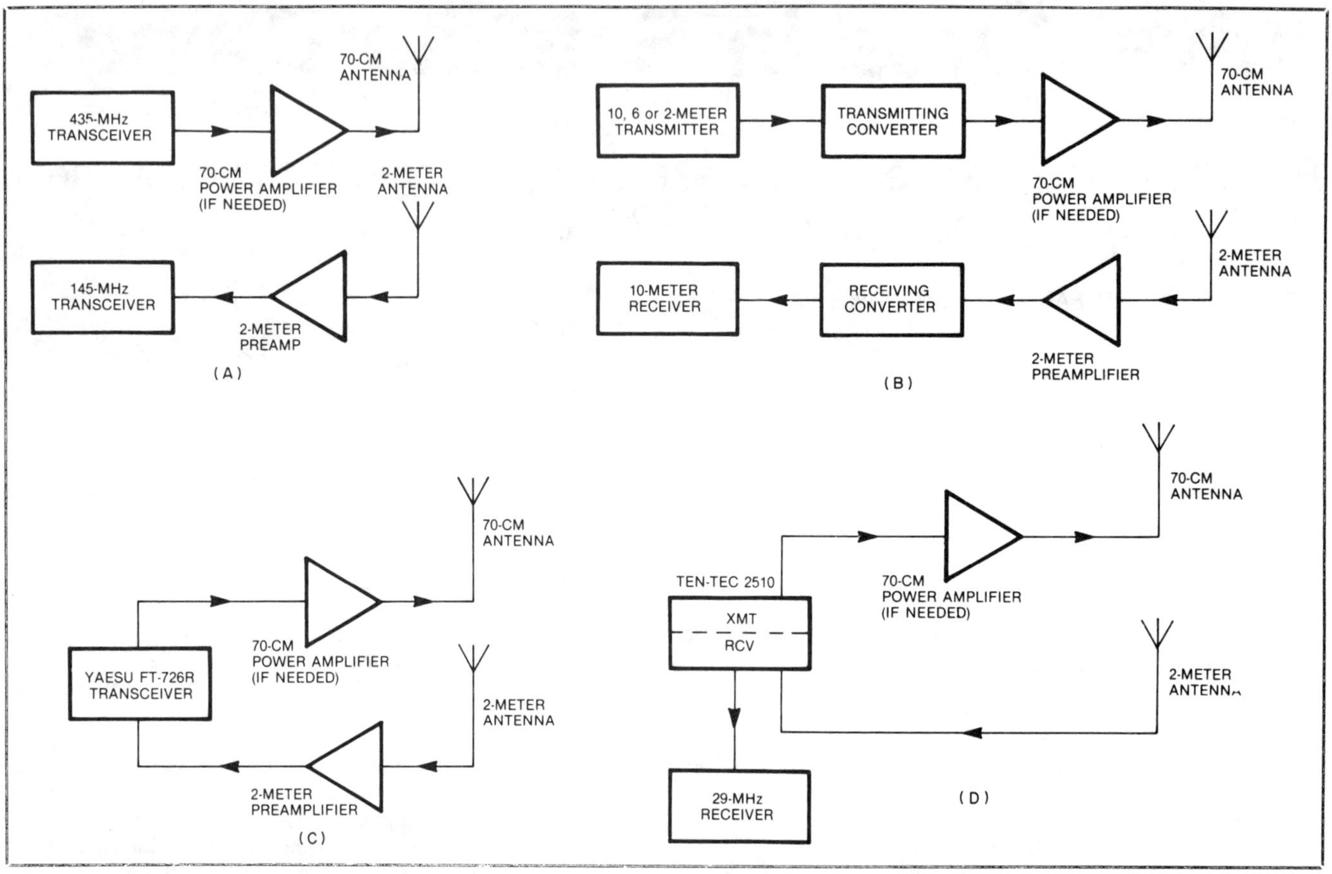

Fig 1—Several different satellite-station configurations are shown here and described in the text. At A, a separate VHF/UHF multimode transceivers are used for transmitting and receiving. The configuration shown at B uses transmitting and receiving converters or transverters with HF equipment. At C, the Yaesu FT-726R can perform both transmitting and receiving functions, full duplex, in one package. The Ten-Tec 2510 shown at D contains a 435-MHz transmitter and a 2-meter to 10-meter receiving converter.

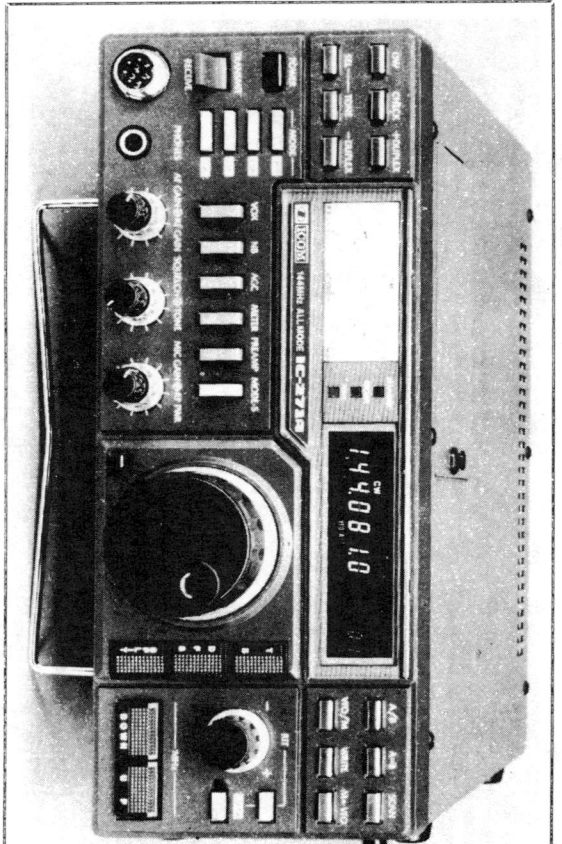

The ICOM IC-271A is typical of the 2-meter multimode transceivers on the market today. It can be used for reception on the OSCAR 10 downlink. The IC-471A, virtually identical in appearance, offers a matching 70-cm, 25-W signal for the uplink.

The major equipment manufacturers listed in Table 1 all make suitable transceivers. The current crop of base-station rigs includes the Kenwood TS-711A, ICOM multimode transceiver also makes an excellent replacement for an FM-only 2-meter rig.

radio. Gear such as the Kenwood TS-700 series, Yaesu FT-225RD and ICOM IC-251 are still popular. Many of these transceivers have been reviewed in *QST*.[1-5]

An excellent solution to receiving OSCAR 10 on 145.9 MHz can be found in the form of receiving converters used with your quality HF transceiver or receiver. The receiving converter consists of a 2-meter preamplifier, a mixer and a local oscillator. The local oscillator frequency is usually chosen so that 2-meter signals will be converted for reception by any receiver that covers the 10-meter band. In addition, a number of manufacturers offer transverters that include a receiving converter and transmitting converter in the same package.

Receiving converters are available commercially from several suppliers listed in Table 1. For those who enjoy building equipment, *The ARRL Handbook* presents several suitable construction projects in Chapter 31.[6]

There are several advantages to using a receiving converter. Your modern HF transceiver or receiver most likely has excellent frequency stability, a frequency readout in 1-kHz or smaller steps, good

IC-271A and Yaesu FT-726R. There are also several compact multimode radios intended for mobile use that will fill the bill. These include the Yaesu FT-480R, Kenwood TR-9130 and ICOM IC-290H. In addition, there are often good buys on the used market, if you're interested in an older

inadequate for AO-10 operation, and that some VHF transceivers do not work well in areas with many strong, nearby signals. If you have QRN problems or live in an area with lots of 2-meter FM repeaters, you may have better luck with a receiving converter than with a VHF transceiver.

Preamplifiers

No discussion of satellite receiving systems would be complete without mentioning preamplifiers. A good, low-noise preamplifier is a great help for receiving those weak downlink signals. Multimode rigs and most transverters will hear much better with the addition of a GaAsFET preamplifier ahead of the receiver front end. While you can add a preamplifier right at the receiver in your station, it may not do you as much good as you think. You'll get much better results if the preamp is mounted at the antenna. Losses in the feed line will degrade the noise figure of even the best preamplifier mounted at the receiver.

Table 1 lists several sources of commercially-built preamplifiers. These are available in several configurations. Some models are designed to be mounted in a receive-only line, for use with a receiving converter or transverter. Others, designed with multimode transceivers in mind, have built-in relays and circuitry that automatically switch the preamplifier out of the antenna line during transmit. Still others are housed, with relays, in weatherproof enclosures that mount right at the antenna. If you like to roll your own, several suitable designs appear in Chapter 31 of *The ARRL Handbook* and in *The Satellite Experimenter's Handbook*, published by the ARRL.[7]

Transmitters

For transmitting to OSCAR 10, you'll need 5-25 W of 435.1-MHz RF at the antenna. This assumes a good antenna, which we'll discuss later. Feed-line losses at 435 MHz are much greater than at HF, so they must be taken into account here. Feed-line losses in a typical installation can easily run 3 dB, so you'll need anywhere between 10 and 50 W output from your transmitter.

Since the possible number of combinations of transmitter power and antenna gain needed to give a satisfactory signal through OSCAR 10 is infinite, satellite users generally talk about their uplink capability in terms of effective radiated power (ERP). ERP takes into account antenna gain, feed-line loss and RF output power. For example, if you run 10 W into a 3-dB-gain antenna, your ERP is 20 W (3 dB greater than, or twice as strong as, 10 W). This assumes no loss in the feed line; all 10 W from the transmitter reaches the antenna. If you use 10 W into a 10-dB-gain antenna, your ERP is 100 W. You could achieve the same 100-W ERP with a 50-W transmitter and a 3-dB-gain antenna.

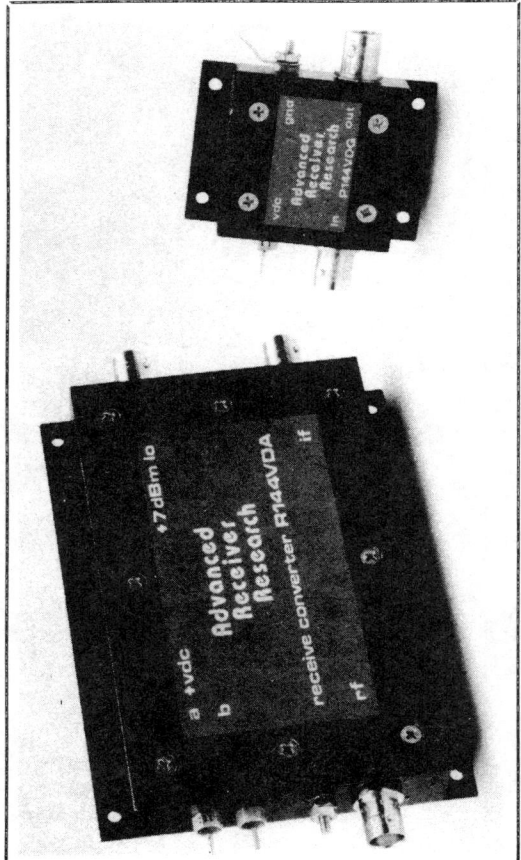

Amateurs who own HF receivers or transceivers and who have no desire for transceive operation on 2 meters may find that a receiving converter such as this Advanced Receiver Research R144VDA will fit their needs. Shown with matching low-noise GaAsFET preamplifier, this unit converts 2-meter OSCAR 10 downlink signals for reception on any receiver that can tune to 29 MHz.

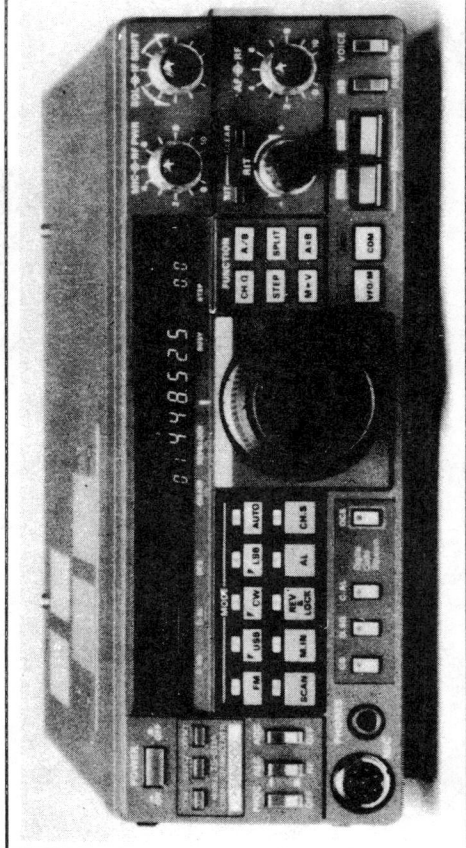

The Kenwood TS-811A is a 70-cm multimode transceiver that can be used to generate a 10-W, 435-MHz signal for the uplink. A similar unit, the TS-711A, may be used for the 2-meter downlink.

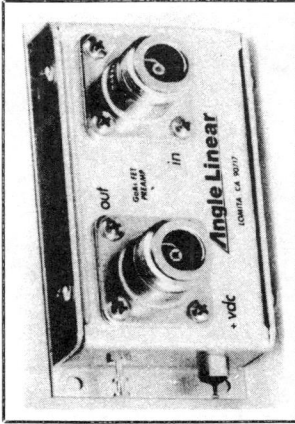

A low-noise GaAsFET preamplifier such as this Angle Linear GaAsFET model is essential if you want to maximize your downlink reception. The preamplifier can be mounted at the antenna for best results.

these features. Cost is another factor. If you already own an HF rig and are not interested in terrestrial 2-meter SSB operation (that is, don't need a 2-meter transmitter), the cost of building or buying a receiving converter will be significantly less than that of even an older multimode transceiver.

The downlink receiver at WD4FAB had for years been a multimode transceiver. Daytime QRN often raised the practical receiver noise floor by 10 to 20 dB, thus eliminating OSCAR 10 daytime communications. Weak downlink signals were no match for the noise. In addition, local FM repeaters could be heard in the satellite passband because the VHF transceiver offered poor rejection of strong nearby signals. Use of a high-dynamic-range receiving converter with a good HF transceiver has, however, solved both of these problems. The lesson here is that many VHF transceiver noise blankers are

SSB and CW crystal filters, an effective noise blanker and high dynamic range. Chances are good that a multimode VHF transceiver will offer some, but not all, of

Stations with an uplink ERP as low as 10 W can be copied through OSCAR 10, but ERP levels of 100 to 400 W are the norm. No matter what your ERP, your signal on the downlink should never be stronger than the AO-10 general beacon at 145.81 MHz. You must have a way to adjust your uplink signal so that it is as strong or weaker than the beacon. We'll return to this point in detail in the operating installment of this series.

If you have a 10-W transmitter, a short run of low-loss feed line and good antenna gain, you're probably all set. Worry no more about added amplifiers. If losses and gains do not add up to enough ERP for you, a 30- to 40-W amplifier may be needed. Some operators have 100-W amplifiers, but with the antennas available today, use of that much power is guaranteed to create an uplink signal that far exceeds the beacon. This is considered by good operators to be an antisocial action. Considerate operators with the 100-W amplifiers quickly reduce drive power to lower the ERP to acceptable levels. Again, use only the RF power that will make your signal no stronger than the OSCAR 10 beacon.

Most satellite operators use UHF multimode transceivers to generate an uplink signal. The manufacturers listed in Table 1 all make 70-cm multimode transceivers that are similar to the 2-meter units described earlier. Although most of these transceivers provide 10-W output, some can deliver 25 W or more.

Unless you are into 70-cm terrestrial communications (and that can be fun, too), there is no need for a complete transceiver. Transmitting converters for use with HF transceivers are available from suppliers listed in Table 1. In addition, some of those manufacturers make transverters that are suitable for satellite work. If you want to try to build your own equipment, *QST* recently presented a 70-cm transmitter construction project.[9]

If you find that you need more 435-MHz power, there are a number of solid-state amplifiers on the market. See Table 1. Choose carefully; you don't need a rock crusher.

Specialty Equipment

Separate transceivers or transmitting and receiving converters are no longer the only way to go. Modern equipment offerings by Yaesu and Ten-Tec, tailored for the satellite user, do it all in one package.

The Yaesu FT-726R starts out as a 2-meter multimode transceiver (see note 2). It is, however, expandable to work on other bands with the addition of optional modules. The Mode-B satellite operator would most likely be interested in an FT-726R with the stock 144-MHz and optional 430-MHz modules. To tie it all together, Yaesu offers an optional satellite module to allow you to transmit on the 435-MHz uplink while receiving on the 145-MHz downlink. This is full duplex operation; the effect is the same as having two separate radios in one box.

Ten-Tec's 2510 is tailored specifically for Mode-B satellite operation.[10,11] This unusual piece of equipment includes a hot receiving converter that converts 145-MHz signals to 10 meters for reception on any HF receiver or transceiver. A low-noise GaAsFET preamplifier is built in, so no external preamp is required. For the uplink, the 2510 has a complete 10-W, 435-MHz SSB and CW transmitter. The 2510 has only one frequency tuning control for the receiver and the transmitter. The receiver automatically tracks the transmitter, an exceptionally useful feature as we'll see when we discuss operating procedures in a later installment.

Antennas

This is probably one of the most controversial areas, as most amateurs are experts and the authors are no exception. The

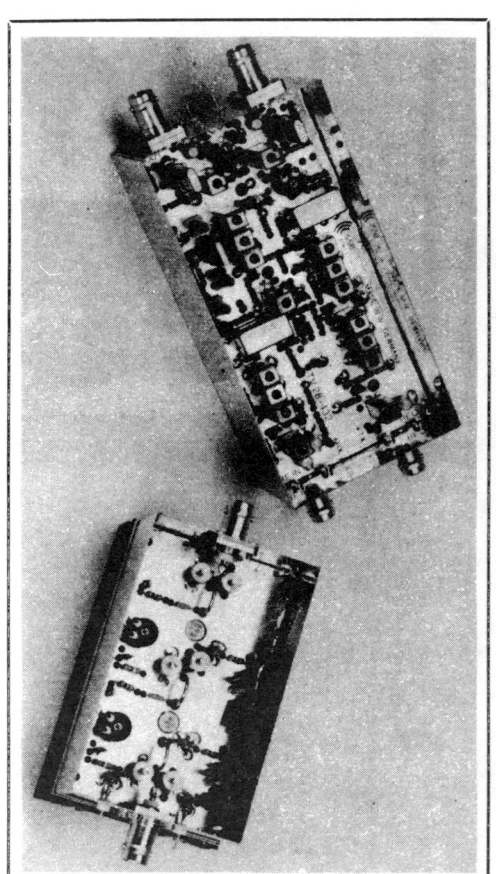

Fig 2 — These helical antennas were used for years at WD4FAB.

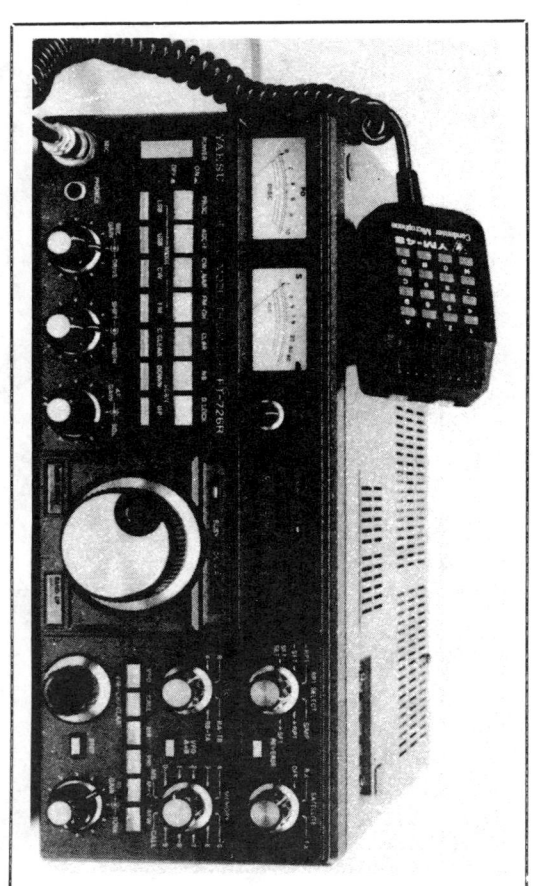

Yaesu's FT-726R is a favorite among satellite users because it can work on both 2 meters and 70 cm. With the optional satellite module that allows full-duplex operation, the effect is practically the same as having two separate transceivers in one box.

Another means of generating a 435-MHz uplink signal is with an HF transceiver and a 10-meter to 70-cm transverter such as this SSB Electronics TV28-432.

Glossary

circular polarization (CP)—describes an electromagnetic wave in which the electric and magnetic fields are rotating. If the electric field vector is rotating in a clockwise sense, as viewed along the path of radiation, then it is called right-hand circular polarization (RHCP). If the electric field vector is rotating in a counterclockwise sense, as viewed along the path of radiation, then it is called left-hand circular polarization (LHCP).

effective radiated power (ERP)—a measure of the power radiated from an antenna system. ERP takes into account transmitter output power, feed-line losses and other system losses, and antenna gain as compared to a dipole.

GaAsFET preamplifier—a low-noise receiving preamplifier that uses a gallium arsenide field-effect transistor as the active device.

The Ten-Tec 2510 Mode-B satellite station is designed specifically for Mode-B operation. The box contains a 435-MHz SSB and CW transmitter, as well as a 145- to 28-MHz receiving converter with a low noise front end.

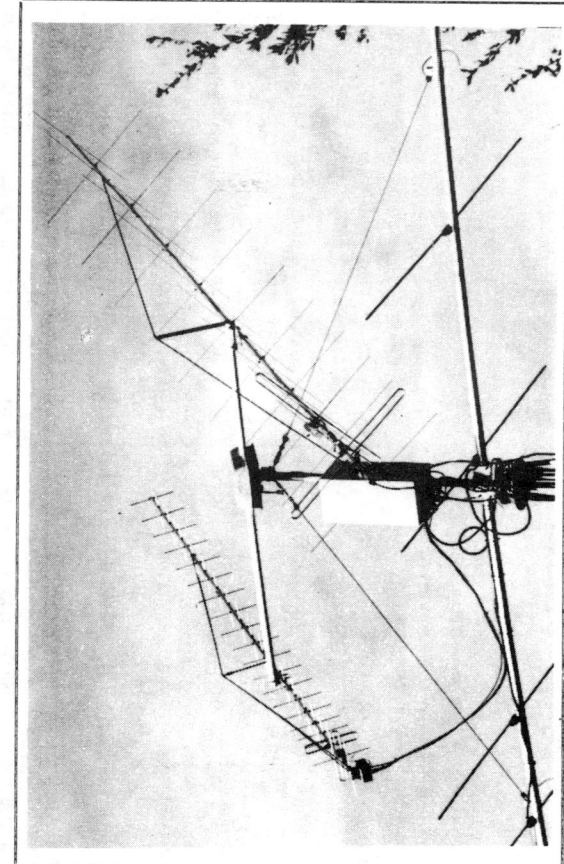

Fig 3—The present satellite array at WD4FAB uses KLM Yagis for 2 meters and 70 cm. The large box on the mast contains a 2-meter preamplifier and a 70-cm power amplifier, as well as power-supply circuitry.

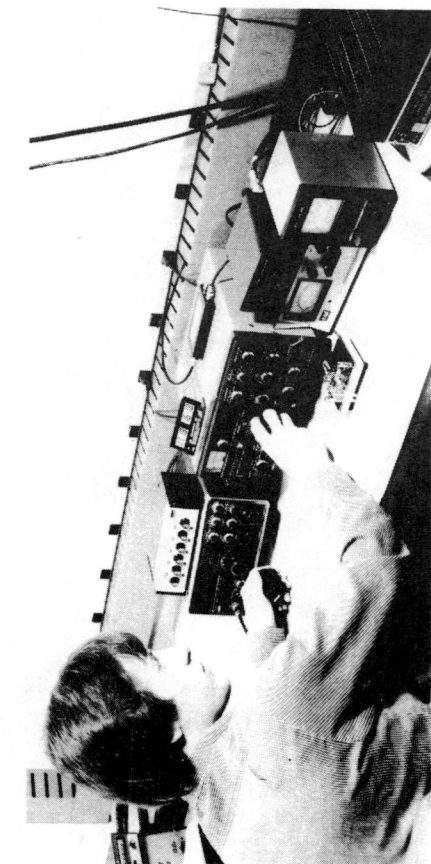

W1INF, the ARRL Laboratory station, has OSCAR 10 Mode-B capability. The uplink is a 2-meter multimode transceiver driving a 2-meter to 70-cm transverter, while the downlink is a receiving converter and 10-meter receiver. A GaAsFET preamplifier is mounted at the antenna.

information presented here is based on years of experimentation at WD4FAB, but it should not be considered the final word.

The best antennas for OSCAR 10 have circular polarization (CP), rather than horizontal or vertical. For years, helical antennas like the one shown in Fig 2 were the way to go for satellite work.[12-14] An eight-turn helical for 70 cm and a huge six-turn helical for 2 meters provided excellent results for OSCAR 8 operation at WD4FAB.[15] For OSCAR 10, however, more gain was needed.

The present satellite array at WD4FAB uses a different method to achieve circular polarization. These antennas, shown in Fig 3, are essentially two complete Yagis mounted perpendicular to each other on the same boom. One set of elements is mounted ¼ wavelength ahead of the other, and the antennas are fed in phase. These particular antennas, manufactured by KLM, have proved to be excellent performers.[16] Cushcraft and Telex/Hy-Gain also manufacture crossed-Yagi satellite antennas.

Perhaps the most significant factor of the KLM "crossed Yagi" antennas is that they are switchable from right-hand circularly polarized (RHCP) to left-hand circularly polarized (LHCP). With AO-10, switchability is important. The side lobes of the AO-10 antenna patterns are LHCP, even though the main lobes are RHCP, and there are substantial side lobe operating periods for any OSCAR 10 orbit. Successful AO-10 operation requires not only antennas with circular polarization, but antennas that have switchable circularity as well.

Satellite antennas should be mounted as close to the station as possible. Height

Satellite Anthology 17

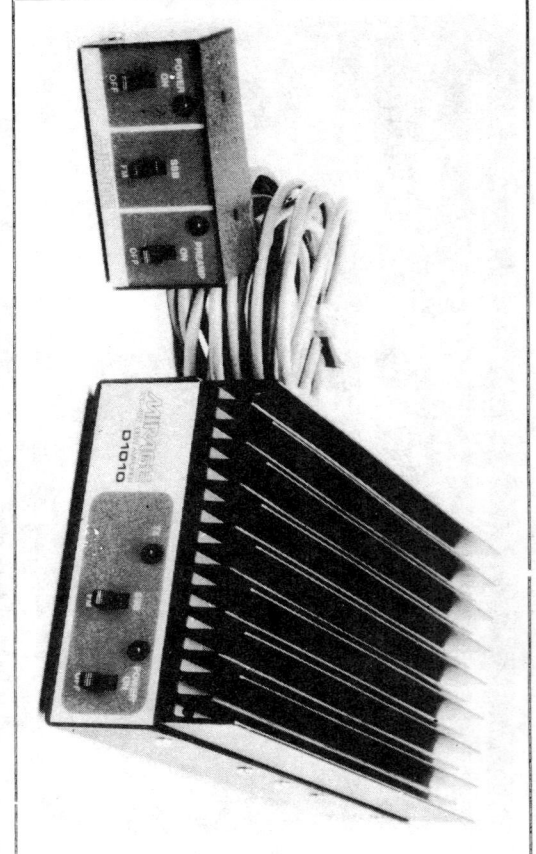

Usually, a little more than 10-W uplink power is required for a good downlink signal. Solid-state "brick" amplifiers such as this Mirage D1010 provide the extra power needed. Care must be taken, however, to use only the minimum power necessary to maintain reliable communications.

above ground makes no difference for satellite work, except that the antennas must be mounted high enough that trees and other obstructions do not block the view of the satellite at low elevations. A low mount allows use of shorter feed lines (lower losses) and often reduces QRN pickup by the antennas. Many operators are able to set up their antennas on a 10- to 15-foot mast right next to the shack and have only 20 feet of feed line. The antennas at WD4FAB are mounted 63 feet above the ground to clear trees, and they require 80 feet of feed line. For feed lines, plan to use good-quality, low-loss coaxial cable from the start, such as Belden 9913. Even better is a run of Hardline with Belden 9913 for the flexible pieces at each end.

We've given you plenty to think about for now. If you need more information, see "A Survey of OSCAR 10 Station Equipment" in a past issue of AMSAT's *Orbit* magazine.[17] Next month, we'll look at useful accessories and antenna rotators, and show you how to assemble all of the pieces into a working OSCAR 10, Mode-B satellite station.

Notes

[1]M. Wilson, "ICOM IC-271A 2-Meter Multimode Transceiver," *QST*, May 1985, pp 40-41.
[2]M. Wilson, "Yaesu Electronics Corp. FT-726R VHF/UHF Transceiver," *QST*, May 1984, pp 40-42.
[3]M. Wilson, "Yaesu FT-480R 2-Meter Multimode Transceiver," *QST*, Oct 1981, pp 46-47.
[4]J. Kleinman, "ICOM IC-290H All-Mode 2-Meter Transceiver," *QST*, May 1983, pp 36-37.
[5]D. DeMaw, "Trio-Kenwood TS-700S 2-Meter Transceiver," *QST*, Feb 1978, pp 31-32.
[6]M. Wilson, ed., *The 1986 ARRL Handbook* (Newington: ARRL, 1985) Available from your local radio store or from ARRL for $18 ($19 outside US). Add $2.50 ($3.50 UPS) per order for shipping and handling.
[7]M. Davidoff, *The Satellite Experimenter's Handbook* (Newington: ARRL, 1985). Available from your local radio store or from ARRL for $10 ($11 outside US). Add $2.50 ($3.50 UPS) per order for shipping and handling.
[8]J. Lindholm, "ICOM IC-471A 70-cm Transceiver," *QST*, Aug 1985, pp 38-39.
[9]J. Reed, "A Simple 435-MHz Transmitter," *QST*, May 1985, pp 14-18.
[10]C. Hutchinson, "TEN-TEC 2510 Mode B Satellite Station," *QST*, Oct 1985, pp 41-43.
[11]D. Ingram, "The Ten-Tec 2510 OSCAR Satellite Station/Converter," *CQ Magazine*, Feb 1985, pp 44-46.
[12]R. Jansson, "70-Cm Satellite Antenna Techniques," *Orbit*, Mar 1980, pp 24-26.
[13]B. Glassmeyer, "Circular Polarization and OSCAR Communications," *QST*, May 1980, pp 11-15.
[14]C. Richards, "The Chopstick Helical," *Orbit*, Jan/Feb 1981, pp 8-9.
[15]R. Jansson, "Helical Antenna Construction for 146 MHz," *Orbit*, May/June 1981, pp 12-15.
[16]R. Jansson, "KLM 2M-22C and KLM 435-40CX Yagi Antennas," *QST*, Oct 1985, pp 43-44.
[17]H. Winard and R. Soderman, "A Survey of OSCAR Station Equipment," *Orbit*, Nov/Dec 1983, pp 13-16 and Mar/Apr 1984, pp 12-16.

Adventures in Satellite DXing

Part 3: Now that you've decided on some equipment for OSCAR work, it's time to assemble it into an effective station.

By Dick Jansson, WD4FAB
1130 Willowbrook Trail
Maitland, FL 32751

Last month, we talked about different ways of setting up your station for OSCAR 10, Mode-B operation. Now that you have a receiver, transmitter and a pair of antennas, you are probably wondering how to tie these parts together into a working system. In this part of the series on satellite DXing, you will discover those sometimes elusive techniques needed to make your radio equipment come alive with action.

Satellite work, like any other specialized facet of Amateur Radio, requires some specialized knowledge. Having the best equipment does not necessarily guarantee success. Presented here are a number of "hints-and-kinks" type ideas that have made OSCAR 10 operation more satisfying for me. Remember that this is my way of converting a basic receiver, transmitter and a pair of antennas into a fun operating position. My solutions are not the only ones—there are several ways to achieve the needed results.

Some of the items discussed here will provide capabilities beyond that of just operating AO-10, Mode B. They also apply to VHF and UHF terrestrial work. It's only natural: I enjoy terrestrial work as well, so I've equipped my VHF/UHF station with several uses in mind. Design your station to suit your own needs.

An often-heard cry is that there is nothing left to build in Amateur Radio these days. Hogwash! There are lots of useful items to construct, and this can be done without the investment of vast fortunes. Nearly everything you will see here has my own handmade (or modified) label attached. This arena also allows room for customizing a job *your* way. Of course, there are commercially available equivalents for nearly everything, if you prefer that route, but you'll miss a lot of fun.

Antenna Rotators

Unlike stations located on the surface of this good Earth, OSCAR 10 will be found

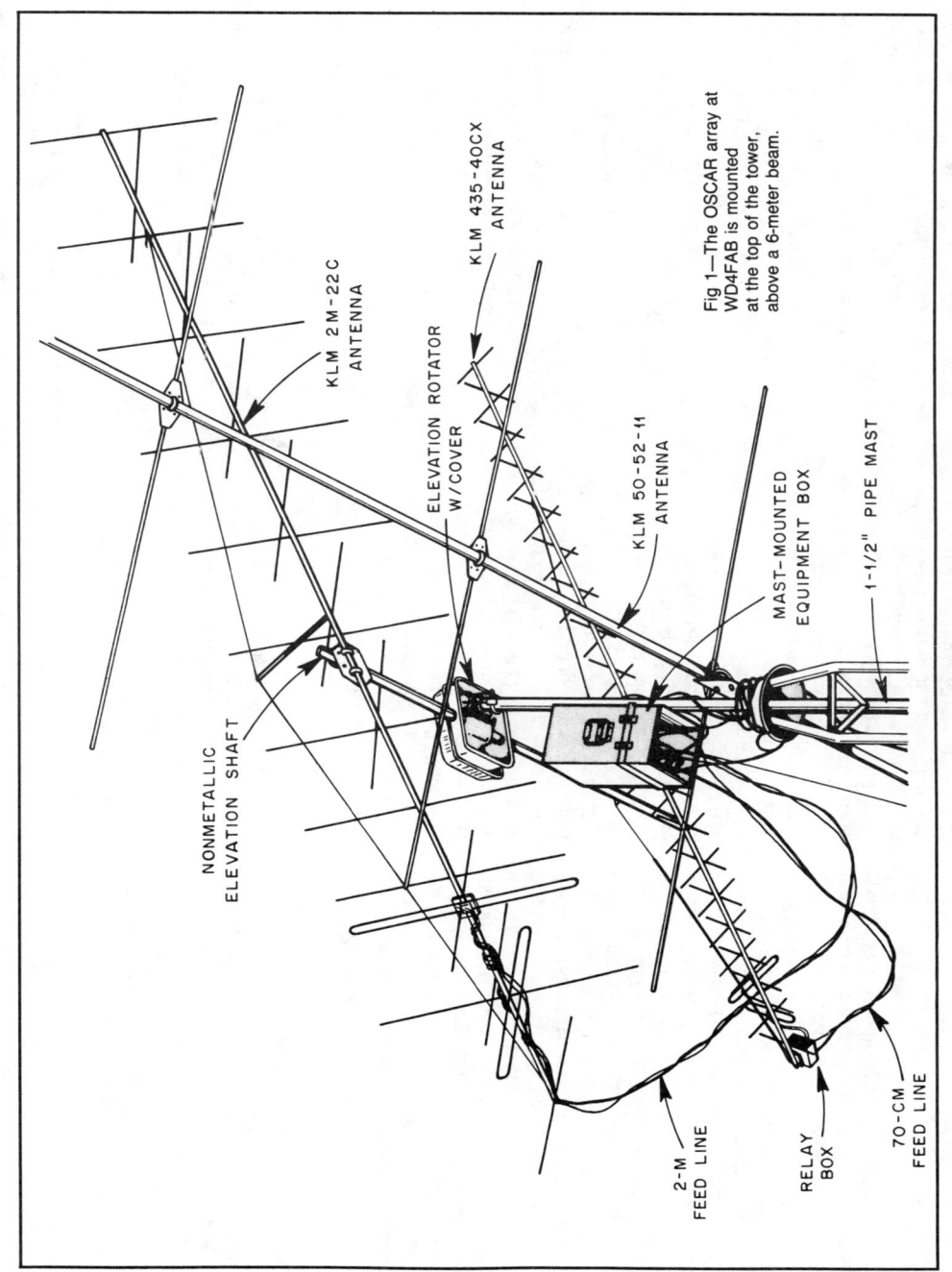

Fig 1—The OSCAR array at WD4FAB is mounted at the top of the tower, above a 6-meter beam.

Fig 2—The elevation rotator, an Alliance U110, is protected from the elements by plastic and aluminum covers. The large white box holds tower-mounted equipment.

Fig 3—Close-up of the U110 showing how it mounts to the mast. Note the PVC pipe that slides over the steel stub protruding from the rotator.

somewhere in the sky above us. You are used to pointing your antenna toward another station by changing the pointing angle, or azimuth (sometimes called az). To find OSCAR 10, you'll also need to be able to control antenna elevation (el). Your antenna must be able to rotate from side to side and up and down simultaneously. See Fig 1. While I will talk about the use of electrically controlled antenna rotators here, you might note that OSCAR 10's motions are slow enough that hand-operated, "armstrong" antenna control is feasible. At times, the antennas don't need to be repositioned for periods of up to four hours.

Azimuth Rotators

Azimuth rotators are common—you've probably got one turning your HF or VHF antenna right now. Antennas for OSCAR 8 and other low-orbit satellites were on the smaller and lighter side, so light-duty TV-antenna rotators such as those sold by Alliance, Channel Master, Radio Shack and others could be used for the azimuth rotator. Today's high-gain satellite array, such as the one described in Part 2 of this series, is a bit large for these light-duty rotators. You really should look for something more robust, such as a rotator recommended for turning a small HF beam or VHF array. Various models manufactured by Alliance, Daiwa, Kenpro, Telex and others are routinely advertised in *QST*.

Elevation Rotators

Elevation rotator selection is somewhat more limited, but there are some interesting things that can be done. Commercially manufactured models are available. The Kenpro KR500, designed specifically for elevating small- to medium-size VHF or UHF arrays, is quite popular among satellite operators. Recent additions to the marketplace include two combined az-el offerings: the Dynetic Systems DR10 and Kenpro KR5400.

A lower cost, commercially manufactured alternative is the Alliance U110 TV-antenna rotator. Rotators of this type have been used by satellite operators (including me) for quite a few years. Despite its relatively light construction, I have had antenna loads weighing up to 80 pounds mounted on a U110! The key to success is to achieve static balance of the antenna mass so that the rotator does not have to elevate a "dead" load. A highly attractive feature of the elevation rotators noted above is that the cross boom to be rotated passes completely through the rotator. This allows you to mount one antenna on each side of center and adjust their respective positions for a side- to-side balanced load.

Figs 2 and 3 show my particular method of mounting the U110. The rotator is clamped to the mast (the one that the azimuth rotator turns) with a plate that permits it to mount 90° from its normal orientation. With the rotator mounted in this position, it is not protected from rain or snow as well as it is in the normal position. I added a cover (an appropriately sized plastic dishpan or bucket is ideal) to afford protection from the elements. A problem with polyethylene plastics, commonly used in kitchenware, is that solar radiation quickly deteriorates their polymeric structure and causes the plastic to break apart. As shown, I have covered the plastic with an aluminum foil baking pan to provide some protection from the sun.

There are other ways to elevate your OSCAR array, although I have found the method just described to be inexpensive and reliable. An ingenious "tilt rather than twist" concept was described in *Orbit* magazine by Al Zoller, W6OTE.¹ This method uses a modified Alliance HD73 azimuth rotator and appears to be viable, despite some limitations for long-boom antennas.

Cross Boom Construction

One requirement not commonly discussed is that of using a nonmetallic elevation axis boom for antennas that have their boom-to-mast mounting hardware in the center of the boom. A metal cross boom will seriously distort the beam pattern of a circularly-polarized antenna, so it's important to make the cross boom from nonmetallic material. My cross boom is made from a combination of metallic and nonmetallic tubing. For strength and stiffness, I used a short length of steel tubing through the middle of the U110. Thick-walled aluminum tubing would work as well. The steel tubing extends for about 6 inches on each side of the rotator. I then installed nonmetallic masting over the steel stubs.

From a structural standpoint, the best nonmetallic material for this job is glass-epoxy tubing, because its stiffness is

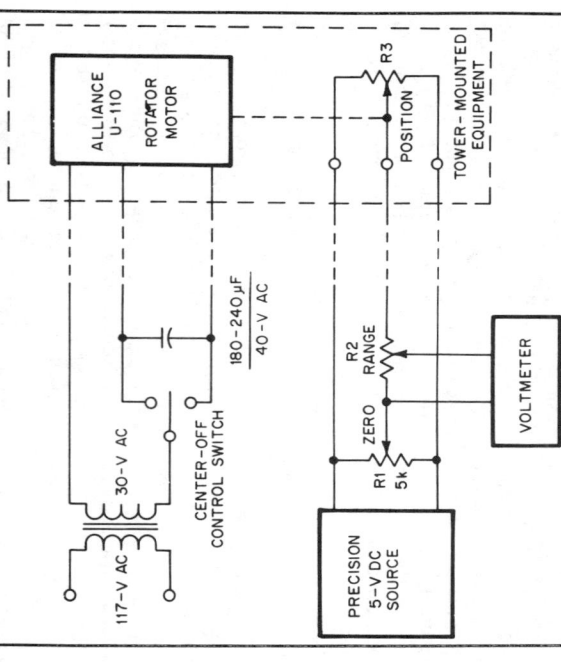

Fig 5—Block diagram of the elevation rotator direction control and position readout.

Fig 6—The circuitry in Fig 5, taken to extremes, provides control and position readout for three rotators.

Fig 4—Details of the position potentiometer mounting and weighted arm.

excellent. You may be able to find this material at an industrial supply house that specializes in plastics. Also, KLM sells lengths of 1½-inch-OD fiberglass masting for this purpose. If you have a rotator that will accept a 1½-inch elevation boom, then your best bet is to use a single section of this tubing.

An alternative nonmetallic material, one that I use, is PVC pipe. Unlike glass-epoxy tubing, however, PVC pipe is not very stiff; it needs help. Based on a suggestion from Nick Laub, WØCA, I built the elevation boom pictured in Figs 1-3. The center piece that fits through the U110 is a 2-foot section of 1.33-inch-OD steel tubing that originally was part of the top support rail of a chain-link fence. Attached to the steel stub on each side of the rotator is a 3- or 4-foot length of 1¼-inch, schedule-40 PVC pipe. PVC pipe is specified by the nominal ID, here 1¼ inches. There are several varieties of 1¼-inch PVC pipe—schedule 40 indicates a thick-wall, heavy-duty version. I was able to slip the PVC pipe over the center stub. The fit is perfect—no machining needed!

Now comes the secret to making PVC pipe capable of supporting satellite Yagis. Insert a wooden dowel into the PVC pipe, along its entire length. The finished dimension of 1-3/8-inch wooden clothes rod dowel (the kind you might hang inside a closet) is just perfect for a slide fit into the pipe. This material is available from most lumber yards. Add a few ¼-inch bolts to each side to secure the pieces, and you've got a sturdy, inexpensive, nonmetallic elevation boom.

Position Indicators

You've probably noticed additional hardware around my elevation rotator. With the use of high-directivity antennas, the accuracy of the U110 control box is questionable. Adding a single-turn, 1-kΩ precision potentiometer provides the ability to closely control and position the elevation boom. See Fig 4. The potentiometer (a large surplus instrumentation model) is attached to the elevation boom with a U bolt and angle bracket. A metal arm with lead weights at the end is attached to the potentiometer shaft. The weighted arm and gravity hold the shaft still while the potentiometer body turns with the elevation boom.

A simple circuit, shown in Fig 5, is all that is needed to control the U110 and to use signals from the precision potentiometer (R3) for an accurate position indicator. R1 is used to zero the scale, while R2 calibrates the range of the indicator. The value of R2 will depend on the value of the voltmeter you use. For example, you might use a 0- to 100-mV meter and adjust it so that 10 mV equals 10° elevation, 30 mV equals 30° elevation, and so on. Just to show that simple circuits can easily be corrupted by some of us, Fig 6 shows a controller for up to three rotators. One *useful* feature of this box is that the voltmeters are OEM digital panel meters. The calibration is set up so that 1 mV equals 1 angular degree.

Tower-Mounted Preamplifiers

Last month we briefly discussed low-noise receiving preamplifiers and said that you'll probably need one to get the most out of your satellite station. For best results, the preamp should be located on the tower or mast, near the antenna so that feed-line losses do not degrade low-noise performance. Feed-line losses ahead of the preamplifier add directly to receiver noise figure. A preamp with a 0.5-dB noise figure won't do you much good if there is 3 dB of feed-line loss between it and the antenna.

Mast mounting of sensitive electronic equipment has been a fact of life for the serious VHF/UHFer for years, although it may seem to be strange or difficult technology for many HF operators. A mast-mounted preamp is not difficult to construct if you prefer to build things yourself. There is a growing number of commercially available models as well. See the list of equipment suppliers in Part 2 of this series. Just to show you that simple ideas can really be made complicated, let's take a look at what I've done with mast mounting of radio equipment.

Take another look at Fig 2. You can't miss the large white box located below the elevation rotator on my antenna stack. Fig 7 shows the interior of this box: It contains a lot of items besides a simple preamp! The box holds two racks of equipment. On the right are two dc voltage regulators with their pass-transistor heat sinks. These regu-

Fig 7—Interior of the tower-mounted equipment rack with the cover removed. The 70-cm equipment is on the left, while power-supply regulators, a 24-cm transmit converter and a 2-meter preamp are mounted on the right.

lators provide on-site-regulated 12-V dc from an unregulated 28-V dc and supply located in the shack. Below the regulators is a 24-cm transmitting converter for Mode-L operation (1269-MHz uplink, 436-MHz downlink). Opposite the regulators on the other rack panel is a 70-cm solid-state power amplifier, a 70-cm preamplifier and relays.

Fig 8 is another view that details the area below the 24-cm transmitting converter. If you look closely, you'll see a 2-meter preamplifier and relays to switch it in and out of the line to the antenna. Before you close the magazine in dismay, remember that this is my particular way of doing things. I'm a fanatic about feed-line loss. There are plenty of successful OSCAR 10 stations that mount only the preamp at the antenna, and that will most likely work for you, too.

Control Circuitry

Fig 9 is a schematic diagram of the control circuitry for the tower-mounted rack. You'll find parts of this diagram helpful, even if you mount just a preamp at the antenna. You should note that I designed this circuit around the surplus coaxial relays that were available at the time. Your version will probably be different and will depend on the relays available to you.

Switching requirements for coaxial relays were the subject of a comprehensive discussion by Joe Reisert, W1JR.[2] Fig 9 is my version of his concepts.

This circuitry performs several functions. For starters, it places the preamp in the line only during receiving periods and takes it out of the line during transmitting periods and at those times when the station is not in use. This is necessary because I use my satellite array for terrestrial transceive operations as well. The circuitry isolates the preamplifier when it is not used for receiving, protecting it from stray electromagnetic pulses (EMP), such as lightning strokes. EMP protection is desirable even if you use the antenna and preamp only for receiving AO-10 signals.

Fig 9 is a bit more complicated than the average mast-mounted preamp setup because I also use 2-meter RF to drive the 24-cm transmitting converter. I have an extra relay (K3) to switch between 2-meter and 24-cm operation. K1, a DPDT transfer relay, switches the antenna and a 50-ohm termination. K2, another DPDT relay, switches the preamp output between a 50-ohm termination and the feed line to the shack. The coaxial cable used for connections between the relays is cut to 0.1 to 0.2 electrical wavelength as recommended by W1JR to

Fig 8—Close-up of the 2-meter preamplifier and relays.

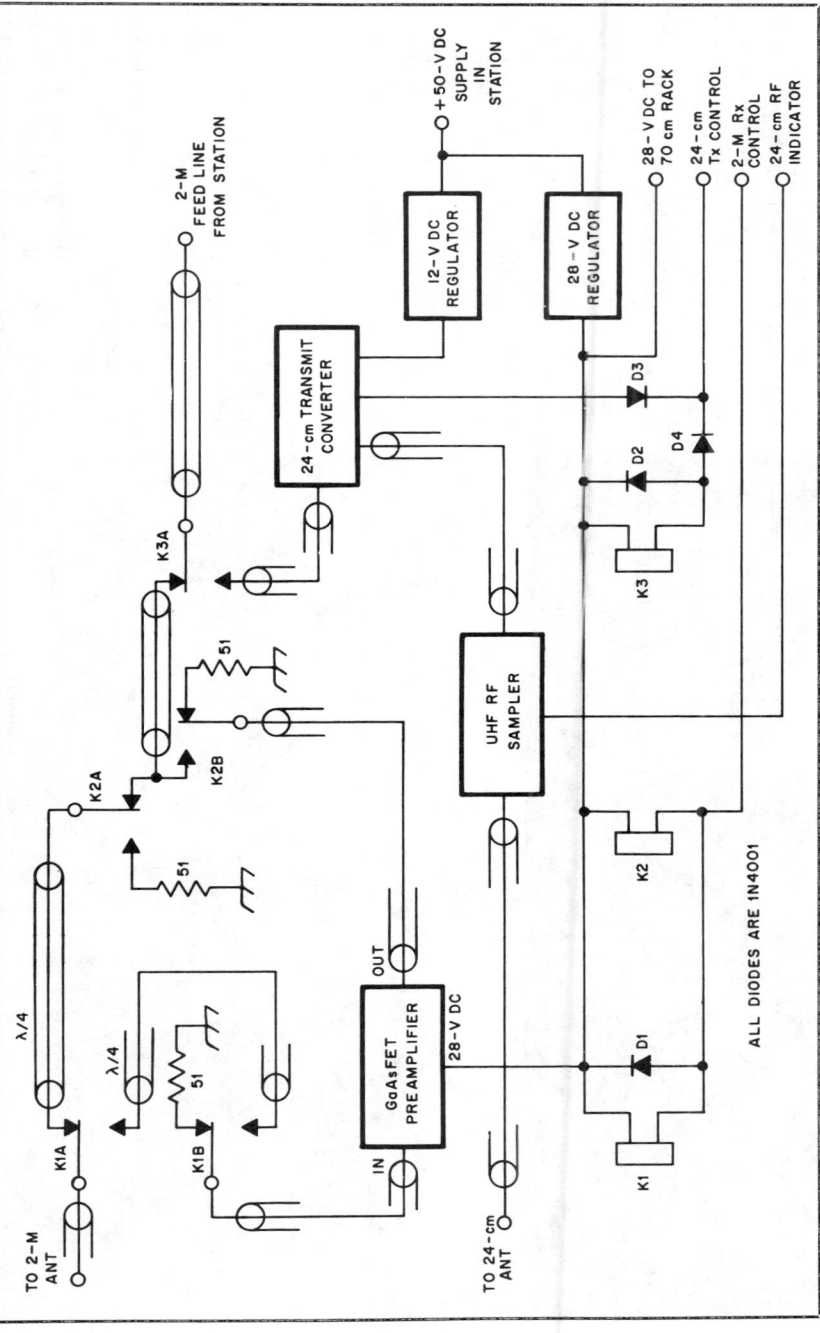

Fig 9—Control circuitry for the mast-mounted 2-meter preamplifier. K1-K3 are surplus coaxial relays.

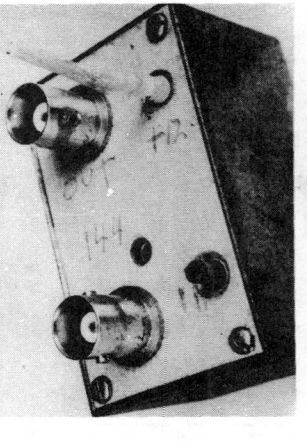

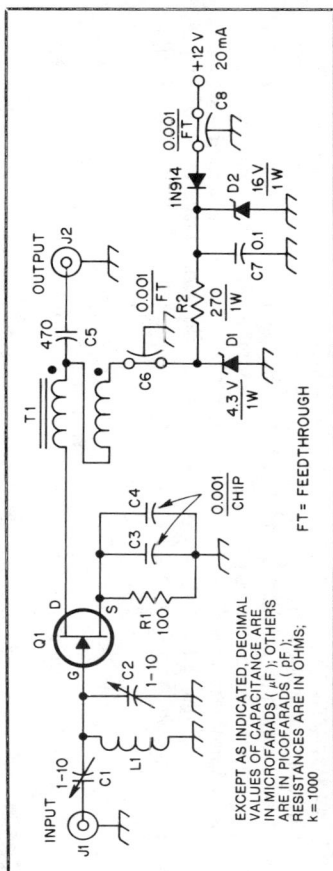

Fig 10—Schematic diagram of the 2-meter, W6PO-type GaAsFET preamplifier. Construction hints may be found in Chapter 31 of *The 1986 ARRL Handbook*.

C1, C2—1-10-pF ceramic or piston trimmer capacitor.
C3, C4—0.001-µF ceramic chip capacitor.
C5—470-pF silver-mica capacitor.
C6, C8—0.001-µF feedthrough capacitor.
C7—0.01-µF disc-ceramic capacitor.
D1—4.3-V, 1-W Zener diode.
D2—16-V, 1-W Zener diode.
J1, J2—Female chassis-mount BNC connector.
L1—6t no. 14 wire, ¼-inch ID, ½ inch long.
Q1—GaAsFET: Suitable parts include MGF1202, MGF1402, NE72089.
R1—100-Ω, ¼- or 1/8-W carbon-composition resistor installed with leads 1/8 inch or less.
R2—270-Ω, 1-W resistor.
T1—12t of a twisted pair of no. 24 enam wire on T37-0 toroid core.

Fig 11—The completed preamplifier is housed in a small aluminum enclosure.

achieve maximum isolation between the transmitted RF and the preamp input. The relays are connected so that they must be energized to place the preamp in line. This setup has worked well for me. I have not lost any GaAsFETs because of EMP or routine RF transmissions, and central Florida is probably the champion lightning-storm area of the country, if not the world.

As you can see, this is all homemade or modified surplus equipment—I enjoy doing things myself. Construction of preamps has been rewarding and relatively easy. I found the basics for the W6PO design (Fig 10) in a newsletter.[3] Similar designs have been documented in *QEX*.[4] There is also a wealth of ideas in Chapter 31 of *The 1986 ARRL Handbook*.[5] The construction process is not terribly complicated, as shown by my 2-meter preamp (Fig 11).

Tower-Mounted Equipment Shelters

A great many amateurs seem apprehensive about placing their valuable radio equipment outdoors. My experiences to date show that such fears are unfounded. Since 1977, I have owned only one Microwave Modules MMt 432/50 (70-cm to 6-meter) transverter. For at least three of those years it was in the wild outdoors of the Florida climate serving its mission well. It has suffered no ill effects other than mild corrosion on the heads of some plated screws that hold the RF connectors. The equipment shown in the photos has been outdoors, without adverse effects, for years.

My present mast-mounted enclosure that you saw earlier is a welded aluminum box I purchased at a local surplus dealer for $6. I used it because it was available and because it is large enough for my needs. You don't need a fancy box like this, especially if all you want to protect is a preamp and two relays.

Fig 12 shows the basic scheme for weatherproofing tower-mounted equipment. This is what I used before I got so carried away with remote mounting. The fundamental concept is to provide a cover to shelter equipment from rain (or snow for you northerners). A 2-inch-deep aluminum cake pan is about the minimum acceptable cover. A trip to the housewares section of the local department store will reveal all manner of plastic and aluminum trays and pans that make great rain covers. As mentioned before, polyethylene plastic must be protected from sunlight. Clear polystyrene refrigerator containers work better, and aluminum is best of all. Choose a cover that is large enough for your equipment; remember to leave room for connecting cables.

You'll notice that the bottom of the rain cover is open to the elements. This is done on purpose and will not cause any problems. Do *not* try to hermetically seal the enclosure. By leaving the bottom open, you provide adequate ventilation, and there will be no accumulation of water condensation. Just make sure that water cannot run into the enclosure by way of cables coming from above. Bend the cables as shown in Fig 12 to provide drip loops.

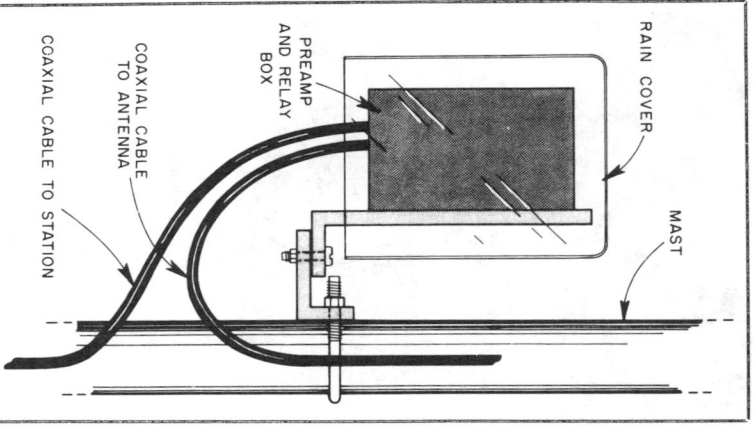

Fig 12—Protection for tower-mounted equipment need not be elaborate. Be sure to dress the cables as shown so that water drips off the cable jacket before it reaches the enclosure.

Transmitting

Fig 13 shows the 70-cm rack from the mast-mounted equipment box. I mounted my 70-cm power amplifier (built on the heat sink at the left of the photo) near the antenna to avoid feed-line losses, but this isn't necessary. I probably wouldn't do it again. Attached to the amplifier output is a coaxial RF sampler for remote power monitoring. To the right of the power amplifier is a 70-cm preamp and coaxial relays I use for OSCAR Mode-L reception and terrestrial operation.

One very important aspect of using GaAsFET preamps with transmitting equipment is getting everything to switch in the right sequence. If you apply voltage to your transmitter, amplifier and antenna relays simultaneously, it's likely that RF will be applied before the relays are fully closed. Such hot switching can easily arc the contacts on your expensive coaxial relays. In addition, if the TR relay is not fully closed, RF may be applied to your preamplifier. Such bursts of RF energy will, in less than the wink of an eye, cleanse your treasured preamplifier of its expensive active device, *guaranteed*. Many pieces of transmitting equipment (especially multimode transceivers) emit a short burst of RF power when switched on or off, so you run the risk of transmitting into your preamp even if you are careful to wait a second to speak or press your CW key.

Ideally, you would set your sequencing up something like this: When you switch into transmit, the coaxial relays change state to remove the preamplifier from the line. Next, the power amplifier is keyed on. The last thing that happens is that the transmitter is enabled. When you switch back to receive, the sequence is just the opposite. First, the transmitter is switched off, then the power amplifier is disabled, and then

Fig 14—The station control unit at WD4FAB fits underneath a transceiver. It houses the TR sequencer and provides instant control of many station functions.

Fig 13—The 70-cm equipment panel holds a power amplifier, preamp and TR relays.

the TR relays change state.

Proper TR sequencing is easy to implement with simple circuitry described in *The 1986 ARRL Handbook*.[6] If you wish to purchase a sequencer, check with the equipment suppliers listed in Part 2 of this series. I've found out the hard way that some form of automatic TR sequencing is necessary with remotely controlled equipment to protect the unwary preamp from cockpit error. Most of us are more fallible than the GaAsFET can stand.

Receiving

The only additional equipment I have found useful applies to those of you who use a receiving converter and an HF receiver for downlink reception. I built an in-line switchable attenuator to use between the converter output and the antenna jack of my 10-meter receiver. I use the attenuator to lower the AGC level and improve the perceived signal-to-noise ratio. In addition, by adjusting the attenuator so the S meter on my HF rig rests at zero, I can give more-accurate signal reports. The attenuator circuit is shown in Chapter 25 of *The 1986 ARRL Handbook*, but I modified it so that there are only three steps: 5, 10 and 20 dB. These three settings allow attenuation in 5-dB steps from 0 to 35 dB.

Station Control

Depending on how complicated you make your satellite setup, you might want to combine most of the switching and control circuitry into a single box so that you have ready access to all controls. Fig 14 shows the system I use. A Minibox cut to a low profile (small enough to fit underneath a transceiver) contains all of the switches I need to control my station, as well as the TR sequencer circuit board. With these switches, I can change polarization on both antennas from RHCP to LHCP; activate, at will, the 2-m and 70-cm preamplifiers; and switch the power amplifier in or out of the line for QRP/QRO operation. I also have the option of manual PTT. The microphone PTT line activates the sequencer.

I hope that this discussion has provided some food for thought for your station. The setup is really not complex; by no means do you need all of the gadgets described here. Perhaps it is all in the mind of the beholder—I happen to enjoy building and modifying equipment. In any event, the last installment of this series will discuss finding and operating through OSCAR 10.

Notes

[1]A. Zoller, "Tilt Rather Than Twist," *Orbit*, Sep/Oct 1983, pp 7-8.
[2]J. Reisert, "VHF/UHF World—Protecting Equipment," *Ham Radio*, Jun 1985, pp 83-87.
[3]C. Osborne, ed., *Southeastern VHF Society Newsletter*, May 1983.
[4]G. Krauss, "VHF and UHF Low Noise Preamplifiers," *QEX*, Dec 1981, pp 3-6.
[5]M. Wilson, ed, *The 1986 ARRL Handbook* (Newington: ARRL, 1985). Available from your local radio store or from ARRL for $18 ($19 outside US). Add $2.50 ($3.50 UPS) per order for shipping and handling.
[6]Sequencing ideas are shown on pp 31-6 to 31-12, 32-37 and 32-38.

Adventures in Satellite DXing

Part 4: At last! You've assembled your station and are eager to work DX. Now it's time to find OSCAR 10 and operate through it.

By Dick Jansson, WD4FAB and Mark Wilson, AA2Z
1130 Willowbrook Trail   Senior Assistant Technical Editor, ARRL
Maitland, FL 32751

In this concluding part of the series on DXing by way of OSCAR 10, we'll examine the subject of locating AO-10 and aiming your antennas toward the satellite. We'll also take a look at those aspects that make OSCAR operation different from routine HF work.

Satellite Tracking

Expert HF DXers know where and when to point their beams, based on years of experience. Tracking satellites is similar in some ways, yet strikingly different in others. It does take experience to become a proficient satellite tracker and to really understand what is going on. In this way, satellite trackers and HF DXers are similar. HF DXers, however, sometimes have to shrug their shoulders at the vagaries of ionospheric propagation, which at times is simply unpredictable. Predicting OSCAR access is much more precise. There is an enormous satisfaction in positioning a simple graphical tracker or dumping a bunch of numbers into a computer and being presented with the information that an object traveling at greater than 18,000 miles per hour is going to pop over your horizon in precisely 38 minutes and 22 seconds. And then it does.

There are two fundamental reasons that you need to keep track of OSCARs. First of all, they move—some fast, others not so fast. You need to keep track of when the satellite is "in view" of your QTH. Second, since most satellite communications require some sort of directional antennas, you need to know where to point the array. Thus, the two primary functions of OSCAR-tracking efforts are position determination and scheduling. There are other functions that might be determined, but these two are the most basic.

Tracking OSCAR satellites requires information in four areas:

1) You need information about the OSCAR to be tracked—its precise location and rates of movement at a precisely defined instant.

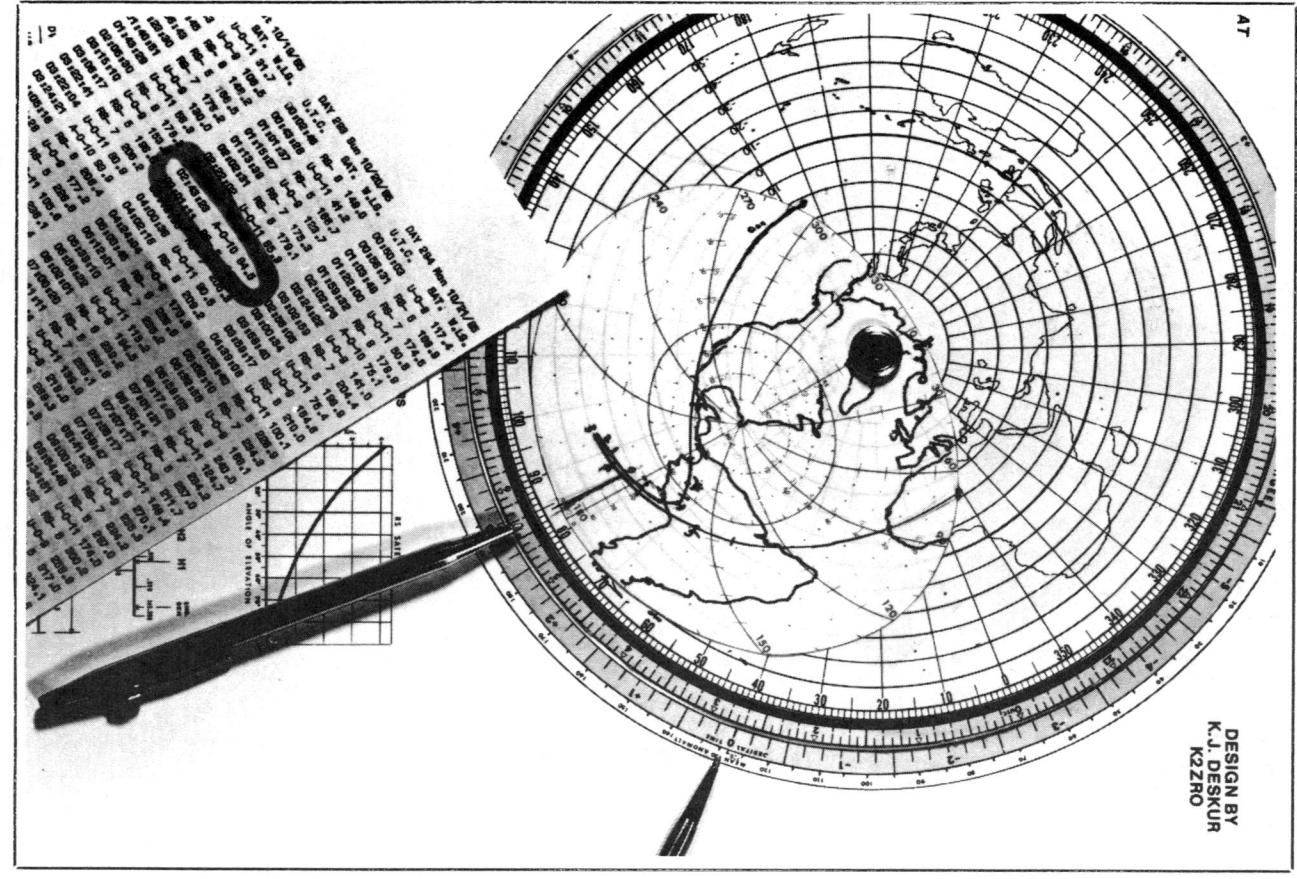

Fig 1—The *Satellipse* from ZRO Technical Products is an easy, inexpensive way to track OSCAR 10. The table at the lower left is part of the monthly orbital information available from ARRL for an SASE.

Fig 2—The OSCARLOCATOR from ARRL contains all of the tools needed to find OSCAR 10. The polar projection map at A is used for tracking all amateur satellites. The clear acetate QTH range circle (B) is centered over your QTH. The AO-10 ground track, shown at C, shows the path the satellite will take. The ground track changes periodically, and updates are published as necessary in QST.

2) You must know your own location to a reasonable degree of accuracy.

3) You must know the time of day reasonably accurately.

4) Most importantly, you need a device to coordinate the first three items. This can be a graphical tracker or a computer program. Both will be discussed here.

Graphical Tracking

Graphical (or manual) tracking methods generally employ a map, typically an azimuthal equidistant projection centered on the North Pole, and one or more clear overlays that allow you to use the map for changing satellite orbits and for different locations on Earth. The overlays allow you to determine which satellite orbits will bring the bird within range of your QTH. They also give beam headings for azimuth and elevation. We'll get to the details shortly.

Most amateur operators who are new to the OSCAR 10 scene are caught up with the idea that they *must* use computer tracking methods to generate the numerical data needed to aim their antennas at the satellite. To those of us who used graphical tracking methods for years to follow the Phase II satellites, such as OSCAR 8, the computer methods were such a revelation that we quickly became married to them. The notion that computers were the only hope was reinforced by the graphical tracking presentations for Phase III satellites given to us at the time of the Phase IIIA demise in 1980. Those manual tracking methods seemed unduly complicated, so we put on our blinders and charged ahead with our computers.

While a great many amateurs do have computers that can be used for tracking (and if you have one, that is the way to go), there are a large number of new satellite operators who are not so equipped. It's easy to be misled into thinking that you must purchase and master a computer before making even a single OSCAR contact. That's enough to make some potential satellite operators lose interest at the onset. Don't be intimidated! Today there are at least two very good, low-cost graphical tracking packages available to you. They give excellent results—finding OSCAR 10 is a snap. Best of all, the investment is downright trivial compared to the cost of even the least-expensive computer.

We highly recommend that you try one of the graphical tracking methods, even if you already have a computer to use for satellite tracking. The introduction to, and

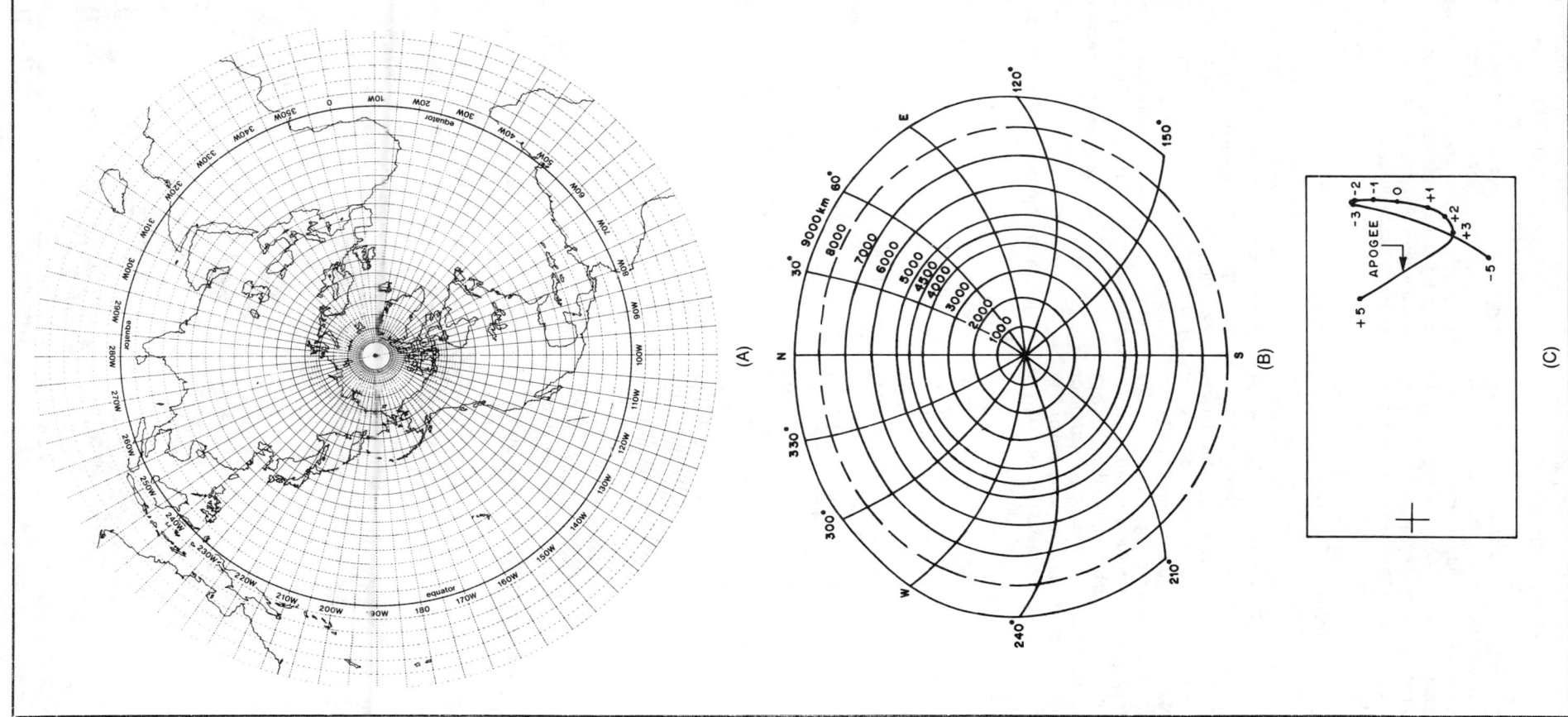

use of, the graphical methods will expand your knowledge and understanding of the nature of the OSCAR 10 orbit and make you a wiser communicator.

The two graphical tracking packages are the *OSCARLOCATOR* from ARRL and the *Satellipse* from ZRO Technical Products.[1,2] See Fig 1. While you are at it, obtain a copy of *The Satellite Experimenter's Handbook* by Martin Davidoff, K2UBC.[3] This publication presents an excellent discussion of graphical tracking. As an added bonus, these publications treat Phase II spacecraft tracking as well as Phase III.

All satellite tracking methods, computer and graphical, need periodic updating of orbital parameters and other reference information. Each satellite has different characteristics, so you'll need data for each satellite of interest. Amateurs can obtain this information from the following sources:[4]

- AMSAT publications, including *Amateur Satellite Report*, a biweekly newsletter.[4]
- *QST*
- Daily W1AW Bulletins[5]
- Project OSCAR Orbital Calendar,[5] a yearly publication of daily satellite reference predictions[6]
- Various AMSAT nets, especially the Tuesday evening net on 3857 kHz at 9 PM Eastern, Central and Mountain times, and 8 PM Pacific time.
- Tabulated satellite reference information, covering all current amateur satellites and good for about six weeks at a time, is available from ARRL HQ. Include a legal-size SASE with two units of First Class postage with your request. Keep a number of SASEs on file, and you will receive the routine updates.

As mentioned before, most graphical tracking methods are based on an azimuthal equidistant projection map of the Earth, centered on the North Pole. A series of clear-plastic overlays is supplied. See Fig 2. Different satellites usually have different orbits, which means that separate overlays are needed for each bird. Both of the graphical tracking packages mentioned here provide overlays for current satellites.

You'll need two overlays for each satellite. The ground-track overlay, which pivots around a rivet positioned at the North Pole, relates the path of the satellite to the map of the Earth. It shows the various locations that the satellite track can take.

The other overlay provides satellite *visibility circles* for your QTH. This overlay tells you which part of an orbit will bring the satellite in view of your QTH. It also shows the azimuth and elevation headings so you can point your antennas at the satellite.

Each of the graphical tracking packages mentioned here comes with complete instructions for use. While it might seem like a lot of work to set one up, graphical trackers really don't require that much effort. They become a lot of fun and a self-satisfying achievement, once the method is learned.

A valid question is, "How good are the graphical tracking methods compared to computer-based systems?" We ran some test cases to find out. In each case, the *Satellipse* results were within 3 degrees of the computer data in both azimuth and elevation. These differences are well inside the half-power beamwidth of highly directive crossed Yagi antennas. Even the most proficient operator would not be able to detect them.

Computer-Based Tracking

For those of you with personal computers, there is a wealth of software available. Most operators are quite satisfied to use the computer to provide numerical data to locate OSCAR 10. They then take this information and aim their antennas accordingly. Changes in antenna positions are infrequent—anywhere from every half hour up to three hours before significant repointing is needed.

Another class of software allows the computer to automatically control antenna position. This approach has some inherent technical problems that are very far afield from computer byte bashing. Unless you are extremely well versed in the software, electronics and mechanics of digitally controlled, closed-loop servo systems, you should forget automatic-antenna tracking. Once you get into tracking AO-10 you will find that you need to adjust the position of your antennas only once or twice an hour by fairly small increments. It is just not worth the effort to control the antenna position automatically! In the days of OSCARs 6, 7 and 8, when the satellite was workable for 16 to 22 minutes each pass, operations in the shack were a bit like the proverbial one-armed paperhanger in a stiff breeze. AO-10 is literally a world of difference.

Satellite-tracking software for a number of computers is available from AMSAT, through the AMSAT Software Exchange (ASE). Most of this software is based on the original work by Dr Tom Clark, W3IWI, that was published in *Orbit*.[7] Since the original work, software specialists have found many innovative ways to express Dr Clark's computational methods. A listing of the various versions as of early 1986 is given in Table 1. For a current program catalog and ordering information, write to AMSAT Software Exchange, PO Box 27, Washington, DC 20044, tel 301-589-6062.

Some commercial software vendors advertise satellite-tracking programs in the ham magazines. An elegant package that is advertised in *QST* is Graftrak II from Silicon Solutions. This program, which operates on the IBM® PC, provides a sophisticated, colorful map display showing the satellite path over your QTH. See Fig 3. Also from Silicon Solutions is Silicon Ephemeris, a satellite-tracking program that has a tabular output. Both packages are available from Silicon Solutions Inc, PO Box 742546, Houston, TX 77274-2546, tel 713-661-8727.

Spectrum West offers Autotrak: Computer Rotor Control for several popular

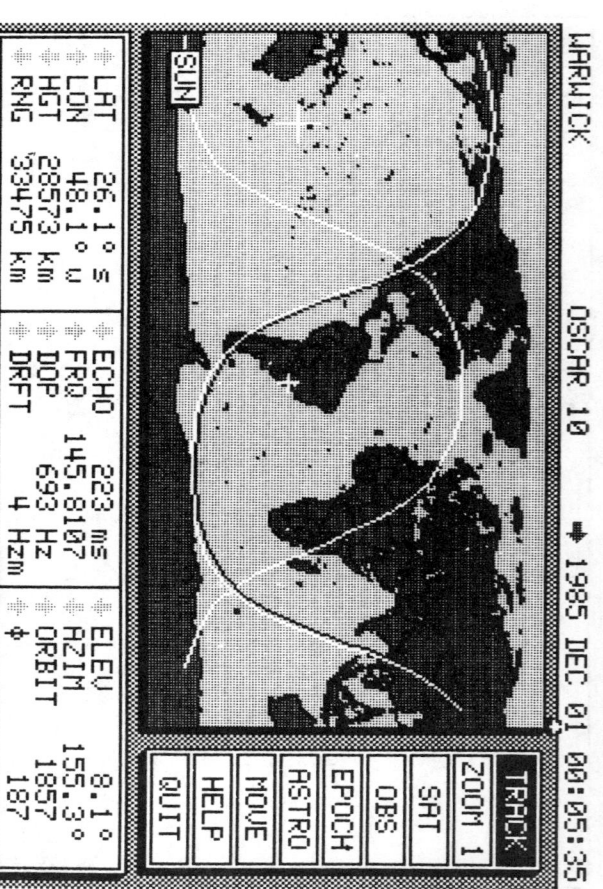

Fig 3—GRAFTRAK II from Silicon Solutions provides an elaborate map display as well as all important satellite parameters. This software runs on the IBM PC and requires an 8087 math coprocessor to help with its intensive calculations.

Table 1
AMSAT Software Exchange Satellite-Tracking Programs

1) Radio Shack TRS-80® Model I, Level II BASIC, 32-kbytes RAM needed. No instructions included; software manual (see item 20) recommended.
2) Radio Shack TRS-80 Model III, 32-kbytes RAM needed.
3) North Star BASIC under North Star DOS for 5¼-inch, hard sector, single- or double-density drives.
4) Microsoft BASIC, version 5.21 under CP/M®, single-density, single-sided 8-inch disk. No instructions included; software manual (see item 20) recommended.
5) Apple® II, APPLESOFT BASIC, on 13- or 16-sectored diskettes or cassette. Menu driven, output to screen or printer.
6) IBM PC, PC-XT or PC-AT version, by W0SL. Menu driven for tabulated output for up to eight satellites in real time. Graphics display of world map. Requires 128-kbytes RAM, DOS 2.0 or later, and BASICA.
7) IBM PCjr version by W0SL. Tracking with graphics. Requires 128 kbytes RAM, DOS 2.0 or later and BASICA as above, but modified to run on PCjr.
8) IBM PC and compatibles version by N4HY. Called QUIKTRAK, it is menu driven for tracking and scheduling and features a new "Window Track" mode for DX.
9) Texas Instruments TI 99/4A, cassette only.
10) Apple II antenna positioning and controlling software by K0RZ.
11) Radio Shack TRS-80 Model 4, for TRSDOS Version 6.0.
12) Radio Shack TRS-80 Color Computer. Requires 32-kbytes RAM and extended BASIC (cassette only).
13) Commodore 64®, AMSAT VR85. Datapoint map of 2000, 20 satellites.
14) QUIKTRAK-2064, enhanced version of AMS-2064, including machine-language file, cassette or disk.
15) Atari® —disk only.
16) Timex-2068, cassette only of W3IWI program.
17) HP-41C programmable calculator, version ORBIT I of the W3IWI program (approximates real-time operation).
18) HP-41C programmable calculator, version ORBIT II of the W3IWI program, converted to run with time module (real-time tracking).
19) Heathkit H89 version of W3IWI program. CP/M version configured for H89, CP/M & MBASIC. Requires 5¼-inch H-17 single-sided, single-density, hard-sector disk.
20) *Using Microcomputer Programs for Radio Amateur Satellite Orbital Prediction* by N5AHD. This manual tells how to use the W3IWI program on Radio Shack, CP/M and S-100 bus computers.
21) UoSAT telemetry capture and decoding software for the IBM PC.

While this computation is somewhat simpler than the W3IWI Keplerian method, it has yet to achieve a great following in the US. For more information, contact the Muellers at 4914 Commonwealth Ave, La Canada, CA 91011, tel 818-790-6695.

Operating Schedule

The last bit of information you need to know before you turn on your gear and start working exotic DX is the OSCAR 10 operating schedule. The satellite does not operate Mode B and Mode L simultaneously all the time. Rather, the transponders are turned on and off according to a fixed operating schedule defined by control stations on Earth. This is done to make sure that the spacecraft's batteries are charged (from solar cells) and discharged (through transponder use) at a rate that will assure the longest possible battery life. The OSCAR 10 schedule needs to be adjusted several times each year because of the satellite's position relative to the sun and earth. The current operating schedule will include the on and off periods for both the Mode B and Mode L translators.

You can obtain the current OSCAR 10 operating schedule from several sources. If you're active on HF, you can hear the schedule on the weekly 75-meter AMSAT net mentioned before. Or, if you've located OSCAR 10, you can get this information on the AO-10 General Beacon frequency of 145.810 MHz. Most of the time, this beacon is sending information about important spacecraft parameters (called *telemetry*) back to Earth by means of a phase-shift-keyed (PSK) signal. PSK telemetry has a raspy buzzing sound. Every 30 minutes, though, the PSK telemetry is interrupted by a five-minute CW bulletin that includes the operating schedule and other bits of important news. These bulletins start on the hour and at 30 minutes past each hour. At 15 and 45 minutes after each hour, another five minutes of information is sent by means of 170-Hz shift, 50-baud RTTY. These RTTY bulletins also contain telemetry samplings, for those interested in that aspect of satellites.

How Well Can You Hear?

This is the point where we put all of your efforts to use. You have high-quality 2-meter receiving capability, modest transmitting power on 435 MHz, good antennas with azimuth and elevation rotators, and a knowledge of where to find OSCAR 10. You *are* ready, aren't you?

It's a good idea to tune to the beacon frequency of 145.81 MHz each time you begin satellite operation. Among other things, the beacon provides a constant signal for peaking your antenna on the satellite. Determine the beacon strength. If you have the receiving capability recommended earlier in this series, you will have useful readings on your S-meter (normally, less than S9 values). It may be helpful to use a signal source and a switchable attenuator to calibrate your S-meter in terms of decibels. This way you can note signal strengths in decibels above the noise, which has more meaning in weak-signal work than normal S-meter readings.

Once you've found the beacon and know that you're hearing OSCAR 10, tune up through the passband (145.830 to 145.970 MHz). Note that the satellite passband is divided according to a voluntary frequency plan, shown in Fig 4. Nearly all of the CW activity is below 145.900 MHz; nearly all of the SSB activity is above that frequency. You will be able to hear packet activity on Special Service Channel (SSC) L2, near 145.840 MHz.

It will take a while to get oriented to Mode B operation because the spacecraft translator is *inverting*, as shown in Fig 5. This means that a signal transmitted at the high end of the 435-MHz uplink passband will come out on the low end of the 145-MHz receiving passband. SSB signals are inverted as well: If you transmit LSB on the uplink, it will come out as USB on the downlink. Common practice is to transmit up to AO-10 on LSB, providing a USB downlink signal.

Table 2
AMSAT-OSCAR 10 Frequency Conversion Chart

Uplink	Mode B ± Doppler Shift Downlink	
435.0323	145.987	EB
435.0423	145.9720	SSC H1
435.0477	145.9620	SSC H2
435.0477	145.9600	GCB UL
435.050	145.9570	ACNF
	145.955	
.055	.950	
.060	.945	
.065	.940	
.070	.935	
.075	.930	
.080	.925	
.085	.920	
.090	.915	
.095	.910	
.100	.905	
435.1037	145.901	
.105	.900	
.110	.895	
.115	.890	
.120	.885	
.125	.880	
.130	.875	
.135	.870	
.140	.865	
.145	.860	
.150	.855	
.155	.850	
.160	.845	GCB LL
435.1647	145.840	SSC 12
435.1747	145.830	SSC L1
	145.810	GB

SSC—Special Service Channel
GCB—General Communications Band
ACNF—AMSAT Calling and Net Frequency
EB—Engineering Beacon
GB—General Beacon
LL—Lower Limit
UL—Upper Limit

computers. This software and hardware package allows the computer to control azimuth and elevation rotators, so the antenna positions are updated as the computer recalculates the satellite position. For more information, contact Spectrum West, 5717 NE 56th, Seattle, WA 98105, tel 206-523-6167.

Another package, using a different computational algorithm than W3IWI employed, is available from Manfred, KG6EF, and Gordon Mueller, KB6BPL. The Muellers provide a Sharp PC-1246 pocket computer and a BASIC routine that was conceived by Dr Karl Meinzer, DJ4ZC.

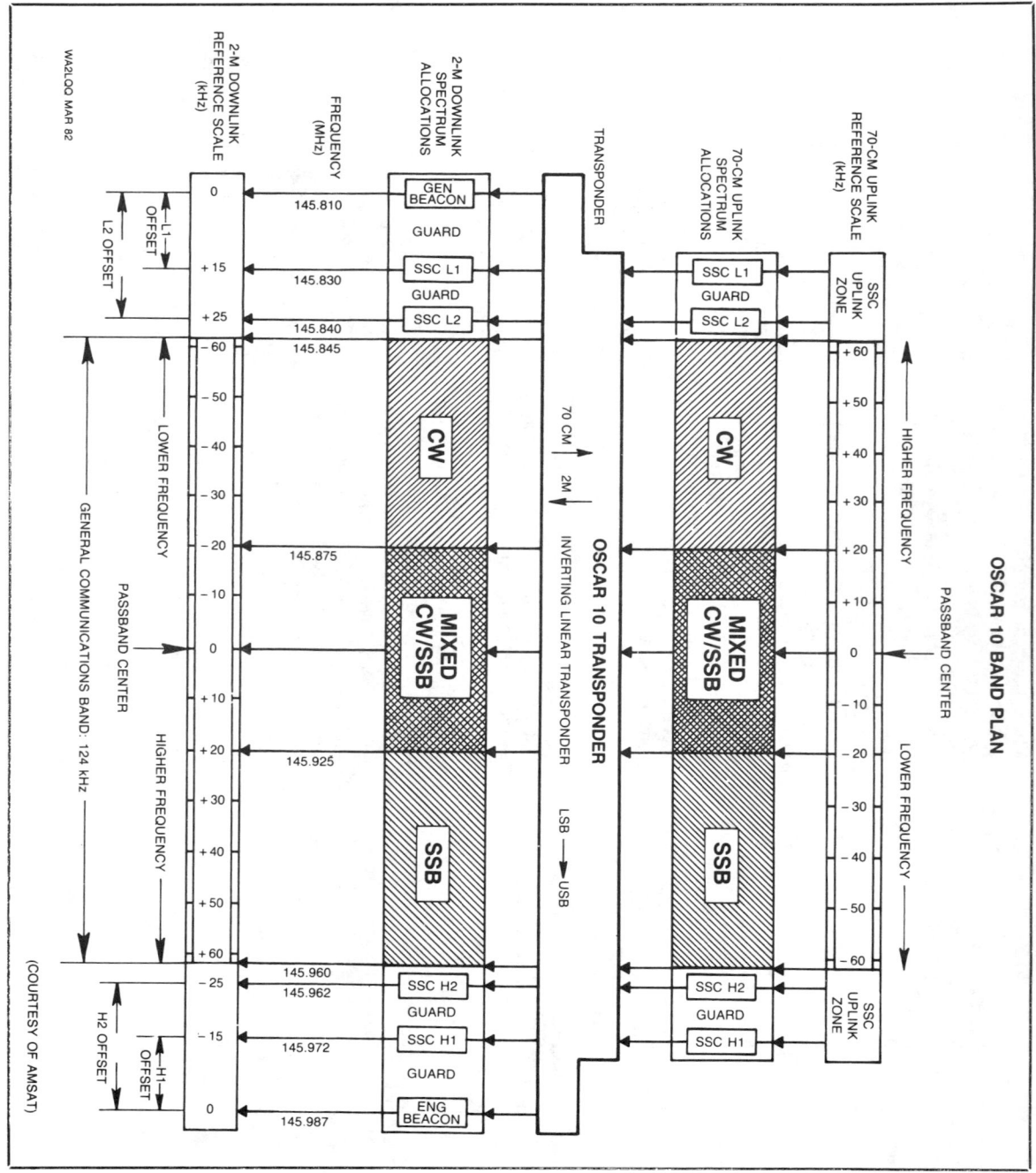

Fig 4—The OSCAR 10 band plan allows for CW only, mixed CW/SSB and SSB only operation. Courteous operators observe this voluntary band plan at all times.

If everything is working right, conversational QSOs can be held with signals ranging from 6 to 15 dB above the noise floor of your receiver. Typically, the beacon is 12 to 15 dB above the noise. Confirmable QSOs have been made with signals as low as 2 to 4 dB above the noise. It is amazingly different from the type of communications that you may have been doing on the HF bands. Note that there is *no excuse* for QRM. Anyone able to receive the satellite should also be able to hear *everyone else* and allow sufficient elbow room for rational QSOs without crowding.

Locating Your Signal

Find some vacant territory on the receiving passband of the satellite for testing your own signal, for example, about 145.960 MHz. There's no need to hurry—AO-10 will be around for awhile. The frequency chart shown in Table 2 will help you to find the right 435-MHz transmitting frequency to correspond with your chosen 145-MHz downlink frequency. If you wish, you can purchase a handy circular slide rule for the uplink-downlink frequency relationships.[8] Assuming you've tuned your receiver to 145.960 MHz, the nominal transmitting frequency is 435.045 MHz. Send a few dits and listen for them on the receiver. Headphones are very helpful here. Tune your *transmitter* frequency a bit on either side of nominal and find your own signal coming back. Note the offset from the nominal frequency; you'll need to know this number any time you want to bring your transmitter to a frequency you're listening on.

The offset is a combination of equipment calibration and *Doppler shift*. Doppler shift is caused by the relative motion between you and the satellite. As the satellite moves toward you, the frequency of the downlink signals will increase slightly. As the satellite passes overhead and moves away from you, the frequency of the downlink signals will be slightly lower than nominal. Doppler shift through a transponder becomes the sum of the Doppler shifts of both the uplink and downlink signals. Since the AO-10 Mode B transponder is inverting, an increase in uplink frequency causes a decrease in downlink frequency, so the Doppler shifts tend to cancel.

If you are using the Ten-Tec 2510 Mode B Satellite Station, tuning the frequency offset is even easier. Set your HF receiver, used as a tunable IF, to 29.0 MHz, and set the 2510 to the desired receiving frequency. Tune the HF receiver a few kilohertz on either side of 29.0 MHz, and you will find your signal. Once set, keep your hands off the HF receiver tuning knob! Adjust the HF rig only for very small Doppler shift corrections and do all of your QSYing with

power among all users. If your signal is louder than the beacon, you'll activate the transponder AGC and degrade performance. It takes only one hog to make communications difficult for everyone.

At Last: Operating Through OSCAR 10

Let's try a CW QSO first. There are two ways of finding someone to work on the satellite, just like on any other band or mode: You can call CQ and hope someone answers, or you can answer a CQ. If you've got a good signal through the satellite, you may want to try calling CQ at first. This way, you won't have to worry about bringing your transmitter to someone else's frequency. You'll be able to get the frequency controls set in advance, allowing you to concentrate on making the QSO. After a QSO or two, venture down the band and try to find another station to call.

Find a clear spot in the CW portion of the passband and bring your transmitter to the frequency. Do this the intelligent way: Look up the frequencies in Table 2, set your transmitter to the right frequency (remember the offset) and send a couple of dots if necessary to zero in on your signal. Resist the temptation to put a brick on the dot lever and crank the transmitter control

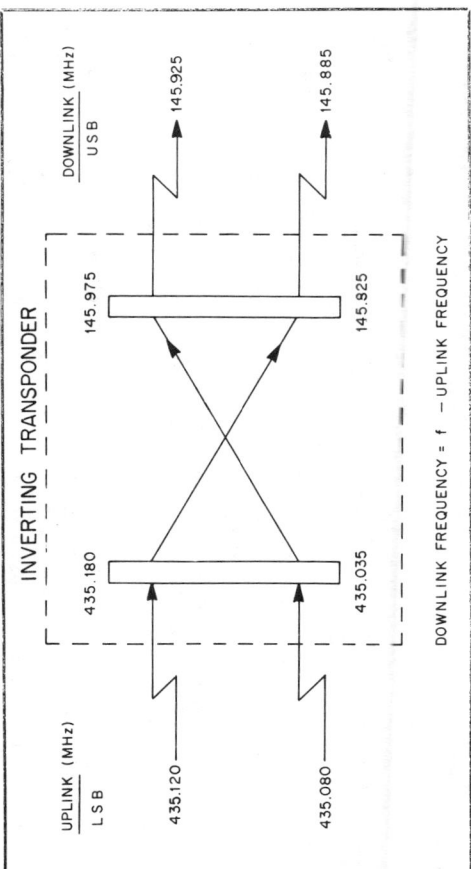

Fig 5—The OSCAR 10, Mode B transponder is inverting. Signals transmitted to the satellite at the high end of the 435-MHz uplink come out at the low end of the 145-MHz downlink. Signals transmitted to the satellite on LSB return to Earth as USB signals. The translation frequency, f_t, is 581.005 MHz.

the 2510 tuning knob. Your transmit signal will always follow your receiver; it's that easy.

Now that you've found your signal, compare its strength with that of the 145.810-MHz general beacon. Your signal strength on the downlink should *never* exceed that of the beacon. If it does, decrease your transmitter power output accordingly. The transponder must share its

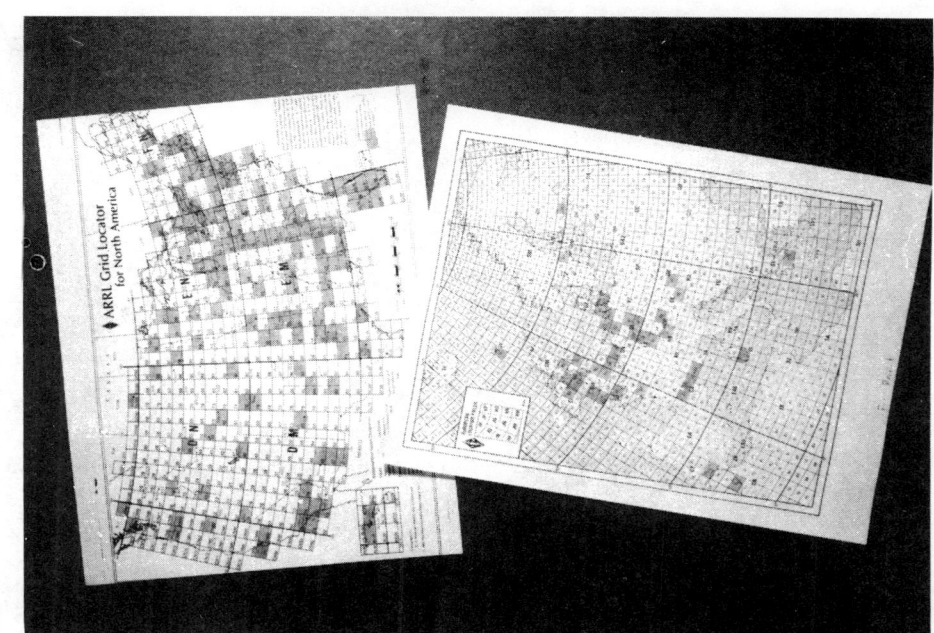

Fig 7—If you're into grid-square chasing, quite a number can be worked via the satellite.

Fig 6—These are but some of the DX QSL cards gathered from Mode B OSCAR contacts at WD4FAB.

until you find your signal. Such "swishing the passband" causes QRM to everyone else and is considered extremely bad manners.

Now call CQ. Although you can hear your signal on the downlink, you'll probably want to use a sidetone oscillator for monitoring your sending. OSCAR 10 is quite a distance away, so it takes a little time for your signal to make the round trip. When AO-10 is in the highest part of its orbit, the delay is about ¼ second. It can be disorienting to listen to a delayed downlink signal while you're sending. It's a good idea to use as a monitor, yet weak enough to let you hear downlink signals between characters or words.

Since satellite operation is full duplex, don't be surprised if someone answers your CQ before you've finished. There's really no reason to wait—it's not like routine single-frequency communications where the receiver is disabled until the end of a transmission. This won't be difficult to get used to, especially if you've used a QSK CW rig before. Should Doppler change during the course of your QSO, tune your transmitter to keep your transmitter and receiver on the same frequency. This way, you won't "walk" down the band as things change. That's all there is to it: Now OSCAR 10 operation is just like what you're used to on HF, only the band's always open when the satellite's in view.

Now Try SSB

SSB operation via OSCAR 10 is similar to CW. The time delay is even more difficult to get over, though. New operators usually stumble over their own return voices when they first try AO-10. You may find it more comfortable to turn down the receiver audio gain control while you're talking. The Ten Tec 2510 has a push-button control to mute the receive audio during transmit. These approaches, while helpful, force the operator back to mono-log transmission without the full-duplex features available from AO-10. With some practice, you should be able to listen to the *pitch* of your return voice and not be bothered by the *content* of what is being said, thus achieving full break-in voice operation.

Although headphones can be used to enhance any weak-signal work, they are mandatory for AO-10 SSB operation. If you don't use headphones, you'll have problems with the receiver audio getting into the transmitter. The result can be anything from lousy, booming transmitted audio to a screeching, full-feedback oscillation through the satellite. Please use your headphones.

An added bonus of lightweight headphones that do not block off all external sounds is that you can hear your own acoustic voice, as well as your voice as transmitted through the satellite. You can readily tune your *transmitter* for the same voice pitch from both sources, thus ensuring that you are on "your frequency" and that any other station that hears you will be able to make contact without chasing you up and down the passband.

The message here is that once you have a QSO established, *hold your receive frequency fixed* and make any Doppler adjustments with your transmitter VFO.

That's about all you need to know to get started on OSCAR 10. There are lots of rare DX stations to chase, if that's what you enjoy. And there is always someone around for a ragchew. You'll find the relaxed pace of AO-10, Mode B operation a relief from the hectic doings on HF. Fig 6 shows some of the QSLs gathered at WD4FAB from Mode B contacts, and Fig 7 shows grid squares worked (both here and abroad).

Good DXing on AO-10.

Notes

[1]The *OSCARLOCATOR* is available from your local radio store or from ARRL for $8.50. Add $2.50 ($3.50 UPS) per order for shipping and handling.
[2]The *Satellipse* is available from ZRO Technical Products, Box 11, Endicott, NY 13760 for $10.
[3]M. Davidoff, *The Satellite Experimenter's Handbook* (Newington: ARRL, 1984). Available from your local radio store or from ARRL for $10 ($11 outside US). Add $2.50 ($3.50 UPS) per order for shipping and handling.
[4]AMSAT, PO Box 27, Washington, DC 20044. Dues are $24 per year. *Amateur Satellite Report* is published biweekly and is included with AMSAT membership.
[5]See the W1AW Schedule in Jun 1986 *QST* for more information.
[6]Available annually for a $10 donation from Project OSCAR, PO Box 1136, Los Altos, CA 98510.
[7]T. Clark, "Basic Orbits," *Orbit*, Mar/Apr 1981, pp 6-11 and 19-20.
[8]The OSCAR 10 "No Ditter" is available from Dave Guimont, Jr, WB6LLO, 5030 July St, San Diego, CA 92110, for $3, postage paid.

OSCAR at 25: The Amateur Space Program Comes of Age

Twenty-five years ago this month, OSCAR I successfully achieved orbit around earth—and amateurs took their first steps into an exciting, new frontier, space communications.

By Jan King, W3GEY, AMSAT VP, Engineering
Vern Riportella, WA2LQQ, AMSAT President
Ralph Wallio, W0RPK, AMSAT VP, Operations
AMSAT, PO Box 27, Washington, DC 20044

In many fields, there are watershed events that mark transitions from one era to another. In aviation, Lindbergh's 1927 solo flight from New York to Paris was such an event. Suddenly, the continents were days closer. Similarly, the 1957 launch of Sputnik partitions history to the pre-Space Age and the Space Age.

In Amateur Radio, the watershed date is December 12, 1961. On that day, OSCAR I was launched. (OSCAR is an acronym for Orbiting Satellite Carrying Amateur Radio.) We predict that in the last decade of this millennium, the significance of that date will come more clearly into focus. It was then, historians of Amateur Radio will note, that the hobby took a sharp turn toward its future: space communications. It was then that ham radio got on track for its major theme in the 21st Century: proliferated networks of hams communicating via multiple media with satellites carrying the bulk of the mostly digital traffic (digitized voice, data and video as a minimum over the amateur equivalent of the Integrated Services Digital Network, ISDN).

On the eve of a quarter century of OSCARs, then, we thought it an appropriate juncture to step back and take the long view. How did we get here? Where are we? Where are we going? And how fast are we getting there?

One way to see where we are going is to chart trends. Mark a few points along a path and soon enough a trend can be discerned. For example, we can classify OSCAR mission complexity and operating environment into three, soon to be four, phases:

Phase 1—Short-lived beacon and/or transponder-equipped spacecraft, from OSCAR I through OSCAR 5, and the recent Russian ISKRA series, which were manually deployed from the Salyut space station.

Phase 2—Longer-lived, multitransponder or scientific spacecraft in low earth orbits, including OSCARs 6, 7 and 8, UoSAT-OSCARs 9 and 11, Fuji-OSCAR 12 and RS-1 through RS-8. (Many new satellites will be added in this class in the future since the low altitude often means strong signals.)

Phase 3—Longer-lived, multitransponder spacecraft in elliptical orbits, including OSCAR 10, Phase 3C (to be launched in 1987) and perhaps Phase 3D in 1990. Benefits of long duration of visibility are offset by complex tracking task.

Phase 4—Very-long-lived multitransponder, multimission geosynchronous spacecraft serving large regions of the earth. Currently undergoing serious study aimed at commencing general use in less than five years.

However, these coarse classifications don't speak to advancements in many engineering and operational areas that have gradually built on the past to produce the capabilities we enjoy today and will enjoy tomorrow. Improvements in power systems, function control, attitude control, telemetry systems and transponder operating frequencies and bandwidth all point toward an astonishing capacity in tomorrow's Phase 4 program. (Refer to Table 1.)

OSCARs I-IV were built by Project OSCAR of California. Australis-OSCAR 5 was built by students in Australia and was launched by NASA on a "ride" arranged

Table 1
Capabilities Growth Comparison of OSCAR

OSCAR	Power	Function Control	Attitude Control	Telemetry	Beacon Transponder
I	Mercury battery	None	None	1-channel CW rate	2-m beacon
II	Mercury battery	None	None	1-channel CW rate	2-m beacon
III	Silver-zinc battery (transponder) solar cells 2.5 W & battery (beacon)	None	None	3-channel CW rate and pulse width	2-m/2-m transponder (50 kHz) 2-m beacon
IV	Solar cells 10 W & battery	None	Spin	None	None
5	Manganese alkaline battery	1 ground command; beacon on-off	Spin & passive magnets	7-channels pulse width modulation	10-m, 2-m beacons
6	Solar cells 5.5-W NiCd battery	21 ground commands	Spin & passive magnets	24-channel CW	2-m/10-m transponder (100 kHz) 10-m beacon
7	Solar cells 15-W NiCd battery	70 ground commands	Spin & passive magnets	24-channel CW, 60-channel Baudot	2-m/10-m transponder (150 kHz) 10-m, 2-m, 70-cm, 13-cm beacons
8	Solar cells 15-W NiCd battery	5 ground commands	Spin & passive CW magnets	6-channel CW	2-m/10-m transponder (200 kHz) 70-cm, 13-cm beacons 10-m, 2-m beacons
9	Solar cells 17-W NiCd battery	Onboard computer & ground command	Gravity-gradient boom	105 channels ASCII Baudot synth-voice digital video CW	2-m, 70-cm, 13-cm, 10-GHz. 7, 14, 21, 28-MHz beacons
10	Solar cells 50-W NiCd battery	Onboard computer & ground command	Spin & active magnets	64-channel ASCII Baudot CW	70-cm/2-m transponder 24 cm-70 cm (950 kHz) 2-m, 70-cm beacons
11	Solar cells 25-W NiCd battery	Onboard computer & ground command	Gravity-gradient ASCII Baudot boom & synth-voice active digital video magnets CW	156-channel	2-m, 70-cm, 13-cm beacons
12	Solar cells 8.5-W NiCd battery	Onboard computer & ground command	Spin & passive magnets	52-channel CW 66-channel PSK	2-m/70-cm transponders analog and digital (100 kHz) 70-cm beacon
P3C	Solar cells 50-W dual NiCd batteries [Watts TBS]	Onboard computer & ground command	Spin & active magnets	64-channel ASCII Baudot	70-cm/2-m transponder 24-cm/13-cm transponder (500 kHz) 2-m, 70-cm, 13-cm beacons

by AMSAT. AMSAT built AMSAT-OSCARs 6, 7, 8 and 10 with its affiliated organizations and help from the ARRL (on OSCAR 8). UoSAT-OSCARs 9 and 11 are the products of the University of Surrey, England. Fuji-OSCAR 12 was a joint project of the Japan Amateur Radio Satellite Corporation (JAMSAT), the Japan Amateur Radio League (JARL), the Nippon Electric Company (NEC) and Japan's National Space Agency (NASDA). More on the history of the Amateur Satellite Program can be found in *The Satellite Experimenter's Handbook* (available from ARRL). Additional reading on FO-12 appears in the Oct and Nov 1986 installments of the Amateur Satellite Communications column and in the June,

Aug and Oct 1986 issues of *QEX* (available from ARRL).

Power Systems

The application of solar-cell-driven battery recharging has been the single greatest improvement in OSCAR power-system design. Early projects predated usable solar-cell technology in terms of output, cost and reliability. While OSCAR III's solar cells and associated secondary battery powered the totally separate 2-m beacon for several months, the primary battery powering the transponder was depleted in 16 days. The first application of solar-cell technology that resulted in an extended working life was aboard OSCAR 6, whose Mode A transponder provided service to the Amateur Radio community for 4½ years, beginning October 1972.

The ultimate demise of every OSCAR project until AO-10 has been battery failure. Consequently, the baseline Phase 3 design includes an auxiliary battery, battery-charge regulator and a reliable means of switching between the two. This redundancy has not as yet been required aboard OSCAR 10, but continues as a vital insurance measure in Phase 3C, which is scheduled to fly in 1987.

Functional Control

The first application of ground-command capabilities for tuning the beacon transmitters on and off, flew aboard OSCAR 5. From this beginning, necessary to demonstrate remote-control capabilities to the FCC, hardwired functional control systems grew to accept as many as 70 different ground commands aboard OSCAR 7.

However, the big breakthrough was the successful application of software-driven onboard controllers, which have come to be known as Internal Housekeeping Units (IHU), beginning with OSCAR 9 and as flown on all OSCAR missions since. The IHU concept allows for at least two long-term benefits:

1) The ability to make decisions aboard the spacecraft, independent of ground control; and
2) The ability to upload new software representing better ideas designed after the bird is in orbit.

Attitude Control

The attitude of the spacecraft relative to the earth is important to ensure the best use of onboard antenna patterns and for thermal dynamics. Although spin stabilization was to be a feature of the ill-fated OSCAR IV mission, the OSCAR 5 project was the first to use both spin stabilization and passive magnets successfully. These attitude-control methods were entirely adequate for all transponder-equipped missions until the Phase 3 design. The scientific-studies payloads aboard OSCARs 9 and 11 require a completely different stabilization technique. A gravity-gradient boom is used for UoSATs.

The tri-star Phase 3 design requires active attitude control to respond to changing sun angles. This control is provided in the form of IHU-controlled electromagnets that are pulsed by navigational software as necessary to maneuver spacecraft attitude. Attitude is determined by sun and earth sensors and is processed by the IHU.

Telemetry

The encoding and transmitting of vital spacecraft operating parameters and conditions has evolved from methods undecipherable by anyone but the primary engineering team (as with OSCARs I, II and III) to transmission of telemetry units in CW, Baudot and ASCII codes with conversion tables available to anyone. In the future, we can look forward to transmission of actual engineering values in plain language, which will be especially useful for elementary and secondary educational purposes.

The quantities of parameters has evolved from just one, the internal temperature of OSCARs I and II, to as many as 156, as transmitted by OSCAR 11.

Beacons and Transponders

Amateur bands used for OSCAR missions have moved steadily higher in frequency as allowed by advances in technology and as the lower frequencies became more crowded with terrestrial operations. Through the years, both transponder efficiency and available bandwidth increased dramatically. OSCAR 10's 950 kHz of transponder bandwidth is more than equal to all preceding OSCAR missions combined.

Transition to the Next-Generation Satellites

Development and mastery of all of these areas has been necessary to bring us to the brink of the Phase 4 era. Without the successes achieved in power systems, functional control, attitude control, telemetry systems and transponders, we would not now be in the position to bring the advantages of satellite-borne transponder communications from the domain of the experimenter to the routine use by amateurs in many other facets of ham radio.

As is apparent, growth in OSCAR complexity and capability has been impressive since the humble beginnings in 1961. Despite the often-dramatic performance improvements between satellites within a generation, and even more so between satellite generations, working OSCAR has remained more or less an esoteric art; only about 3% of active US amateurs consider themselves "OSCAR-active."

A surprisingly high number of active US hams have tried OSCAR at least a couple of times. For one reason or another, they found it did not retain their interest; at least not in terms of the effort required to effect

An Air Force Technical Sergeant admires the handiwork that went into designing and building the world's first nongovernmental communications satellite, OSCAR I. While circling earth, the 10-pound satellite transmitted the word "HI" in Morse code.

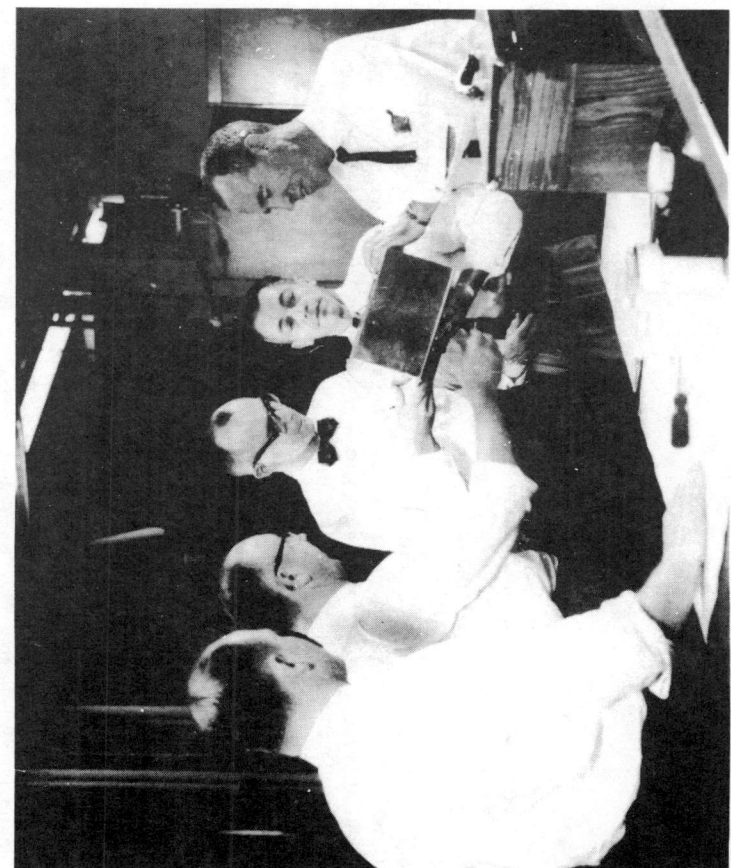

OSCAR I was the brainchild of Project OSCAR, a group of California hams. The members shown here are (l-r) Gail Gangwisch, Nick Marshall, W6OLO, Don Stoner, W6TNS, Chuck Towns, K6LFH, and Fred Hicks, W6EJU. Project OSCAR also was involved in later amateur satellites.

Building today's amateur satellites requires an outstanding range of skills. Here, Dick Daniels, W4PUJ (right), and Wolfgang Mueller (from the German rocket manufacturer, MBB) wear special suits while loading potentially explosive, highly toxic propellant into AMSAT-OSCAR 10 prior to its launch in 1983.

a QSO. And *that* seems to be a major theme with persons we've interviewed regarding their experiences with OSCARs. We've found that about 15% of all current hams have had at least one OSCAR QSO in their ham career. Why this group should outnumber the "regulars" by about 5 to 1 has been the source of protracted soul searching among our future-system architects. "Why," they justifiably ask, "is something so inherently interesting (space communication) such a disappointment in terms of obtaining and maintaining the interest of a large proportion of the amateur population?"

The question is significant in light of plans for our next generation of satellites, Phase 4. For, although the trend lines of past evolutionary growth in satellite capacity and functionality point to several potential growth areas, the consensus among AMSAT long-range planners is that it may be time for *revolutionary* growth instead of evolutionary growth in satellite-system architecture. It may be time for a change in the way we look at satellite systems and how they interact with the general Amateur Radio community.

Moreover, there is a special sense of urgency associated with this introspection. While technology advances have made more OSCAR capacity available to amateurs, it has also sharpened the appetite of commercial interests for the very heart of our hobby: our precious spectrum. The same technology that is making it possible to enhance present and future OSCARs ironically is placing our hold on the VHF/UHF spectrum at risk. Spectrum that was thought useless for commercial purposes years ago is now deeply coveted and eagerly sought by entrepreneurial interests.

The popularity of all prior OSCARs has been throttled by two main factors: access and functionality.

"Access" really means "ease of access" or, alternatively, "convenience." In order to be convenient, an OSCAR needs to appear regularly (at a given time of day), and it needs to be enduring (stick around for long enough for a few QSOs, at least). While AMSAT-OSCAR 10, in its high, elliptical orbit, has improved access in meaningful ways, the major drawback has been that it is not sun-synchronous. That is, its appearance tracked neither with the sun nor human activities (such as work/play schedules), which *are* synchronized with the sun. Nevertheless, AO-10 did provide endurance. It could be in view often for eight- or nine-hour periods, during which thousands of QSOs could transpire.

"Functionality" means, essentially, "What can I do with it?" The conventional wisdom holds that satellites will be truly popular when they can do more than 20 meters can do most of the time. If satellites could do what 20 meters does for less money, that would probably accelerate the popularization process. Well, AO-10 has done some of the things 20 meters does and some of the things it does not. Like 20 meters, AO-10 has provided international coverage. Unlike 20 meters, it has not been choked with QRM. Neither has AO-10 been tied to the sun in terms of when it's on and when it's off. Neither has AO-10 been notably affected by geomagnetic storms or sunspot cycles. But, whereas when 20 meters is very good, signals can be 40 dB over the noise, signals on AO-10 have rarely exceeded 12 dB above the noise. And, spin modulation (QSB) is an effect associated with satellites, not 20 meters.

The point is this: Given the traditional equipment and experience base of the active amateur community, there has been insufficient motivation to become satellite regulars. "What can I do with satellites that I can't do on HF?"

Given this reasonable question, then, let's look at revolutionary ways to provide communications which *can't* be accomplished using available HF techniques. For example, let's provide a way for linking the minimum Amateur Radio station, say a 2-meter hand-held radio, with another hand-held 10,000 miles away through a gateway or teleport in the vicinity of the hand-helds. Let's look at ways of trunking terrestrial packet networks into a global network. Let's consider how we might address emergency voice bulletins to a large portion of the Amateur community through thousands of repeaters across the country using an alert broadcast code and selective addressing from the next-generation OSCARs. Let's see what we can achieve with the latest in digital TV and compression techniques in a new amateur context. Most important, however, let's look at ways of making the next generation of OSCARs truly justify not only themselves in terms of intrinsic merit, but rather in terms of service they can provide to the public. *That* spells revolution, not evolution!

Next month, we'll take a look at the transponders and communications possibilities of the next-generation satellite, Phase 4.

OSCAR at 25: Beginning of a New Era

Easy-to-use space communications may at last be on the way. After 25 years as a technical novelty, AMSAT introduces a new generation of satellites. Public service and accessibility are the watchwords.

By Jan King, W3GEY,
AMSAT VP, Engineering
Vern Riportella, WA2LQQ, AMSAT President
Ralph Wallio, W0RPK, AMSAT VP, Operations
AMSAT, PO Box 27, Washington, DC 20044

Last month, we looked at the beginnings and the development of the Amateur Satellite Program from OSCAR 1 through Phase 3. This month, we'll look at the next-generation satellite, Phase 4, and its various transponders and communications possibilities.

In September 1986, Jan King completed a Phase 4 Engineering Study Plan. In it was depicted a preliminary architecture for a two-satellite geosynchronous system that AMSAT believes could be in operation by 1991-1992. The Plan suggested a one year course of study for Phase 4, during which specialists in various technical fields will look at each facet of the design. The design team will then advance the initial concept to a workable preliminary design. If, after the year, the team feels they have a design that meets the objectives, the AMSAT Board of Directors will be asked to authorize initial construction activities.

Working in parallel with the Phase 4 Engineering Study Team, comprising about two dozen experts, will be two other teams: the Ways and Means Team and the Operations and Applications Team. The Ways and Means Team will be looking into ways of developing resources to enable construction of Phase 4. Besides traditional scouting for donations, gifts and grants, these folks will be looking for donations of key resources (like rare skills), donations in kind (of specific hardware needed), and so forth. They will ferret out the $1-million-plus resources necessary to make this program turn real.

The Operations and Applications Team will be looking at two aspects:

1) How to optimize the strawman satellite architecture to the projected needs and capabilities of terrestrial users of the 1990-2000 time frame.

2) How to prepare the user community for the advent of truly easy-to-use satellite communications.

What will Phase 4 be like? How will it be to use? According to the preliminary (strawman) concept, initially there will be two satellites placed in geosynchronous orbits. The coverage areas (footprints) of each are shown in Figs 1 and 2. AMSTAR East would be positioned over the equator at 46.6° west. (AMSTAR is a preliminary designation for AMSAT's Phase 4 satellites.) From there, it would cover everything east to Helsinki and Durban and west to Seattle. AMSTAR West would cover everything from Boston west to Tokyo and central Australia. Although technically difficult, it might be possible to link the two birds (crosslink) in such a way as to enable a two-satellite QSO from, say, Athens to Melbourne.

What's especially attractive about the geosynchronous orbit is that the old bugaboo about tracking is gone completely! You just set your antenna at a given spot in the sky and, essentially, weld it in place. You never have to move it: no computers, no locators, no nothing; just AMSTAR in the sky 24 hours a day, 365 days a year providing the kind of facility

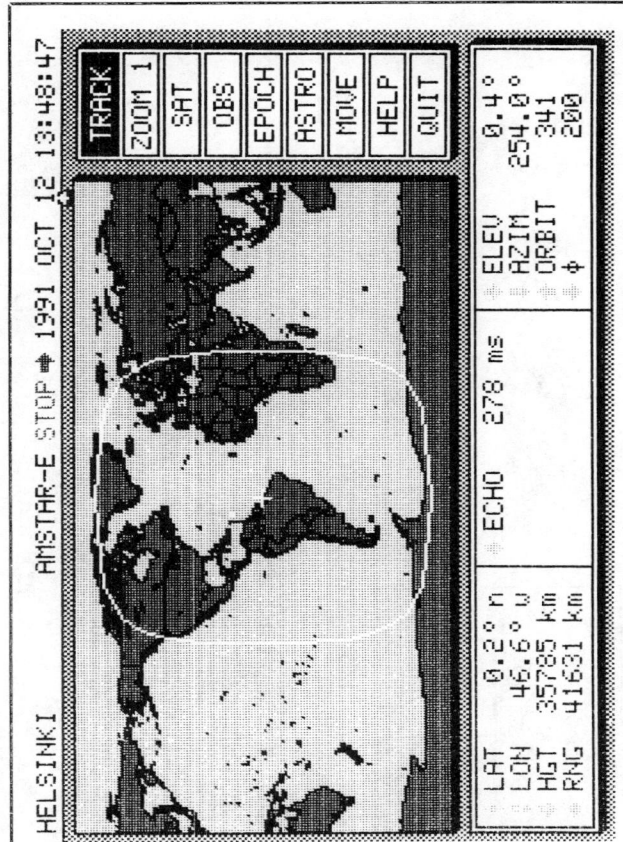

Fig 1—Footprint of AMSTAR East (see text).

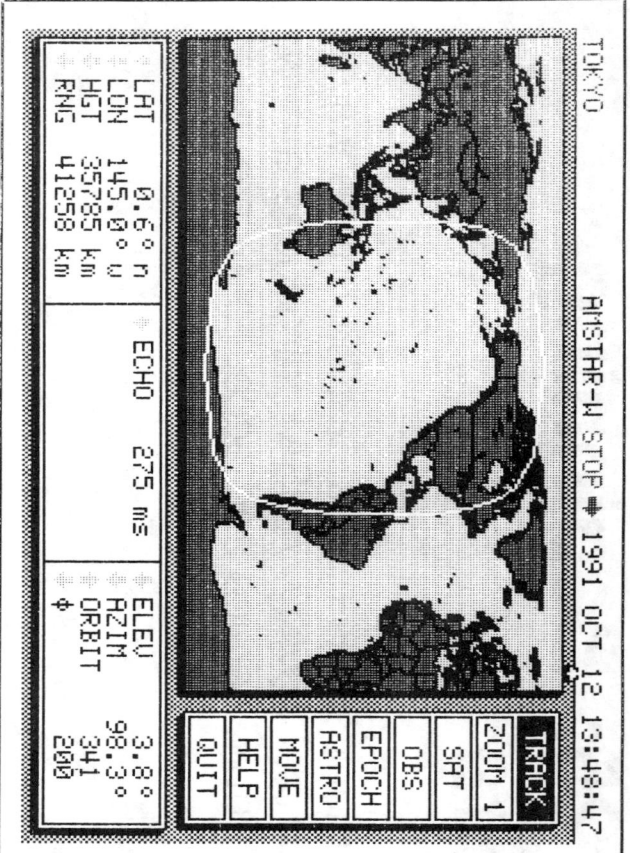

Fig 2—Footprint of AMSTAR West (see text).

a project sponsored by ARRL. Mode J has recently been reborn with its employment on the new Fuji-OSCAR 12 from Japan. As may be seen in Table 2, Mode J involves a 2-meter uplink and a 70-cm downlink. Mode J is especially popular in Japan because intense 2-meter QRM makes reception of the relatively weak 2-meter downlink of, for example, Mode B (70 cm up, 2 m down) very difficult. On the other hand, the 70-cm downlink is not subject to comparable QRM levels in Japan.

Mode L is a relatively new mode, having flown on AMSAT-OSCAR 10 in 1983 for the first time. With 24-cm uplink and 70-cm downlink and fully 800 kHz of bandwidth, it was designed as a safety valve to absorb anticipated user growth on AO-10's Mode B. That growth eventually did reach a stage where it would have likely spurred Mode L use, except that the Mode L transponder developed sensitivity problems. It was infrequently used for communications and occasionally for experimental purposes.

The combined Mode JL will have its first space test next autumn when the latest Phase 3 satellite, Phase 3C, is launched. With Mode JL, 2-meter and 24-cm uplinks each result in 70-cm downlinks. Given the user equipment shown for Mode J in Table 3, the SSB user can expect an average downlink signal-to-noise ratio (S/N) of 10.5 dB (see Table 4). Mode L users do a little better on average with about 11.3 dB

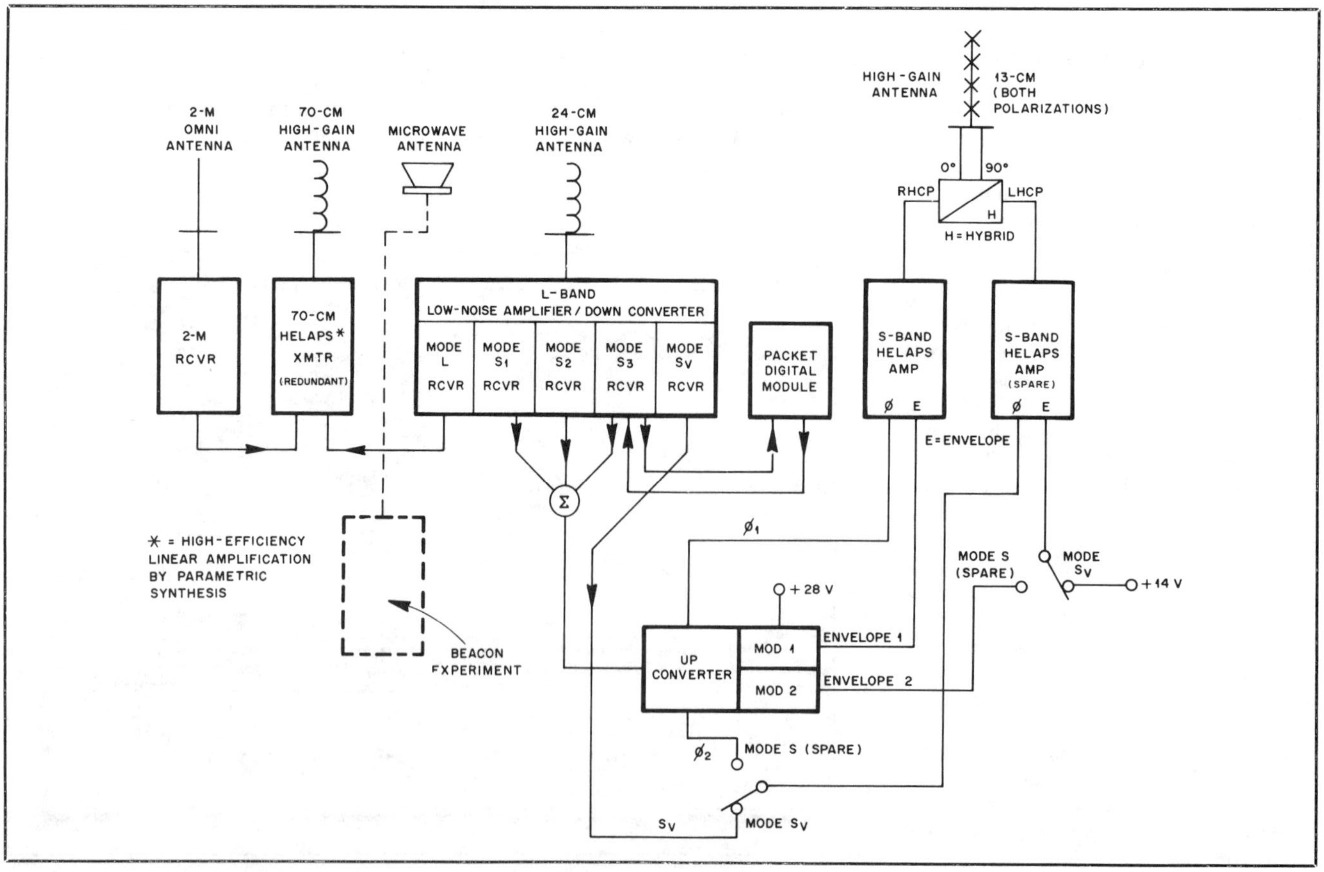

Mode JL.

Mode JL is a combination of two modes (J and L) that have been used previously for OSCARs. Mode J (named for JAMSAT, our Japanese colleagues) first flew aboard AMSAT-OSCAR 8 in 1978 in emergency communicators and ordinary would-be satellite users have been seeking for years.

What kind of communications services might be enabled by Phase 4? Let's look at the various transponders and examine briefly their capabilities (see Fig 3).

Fig 3—Block diagram of Phase 4's transponders and their capabilities (see text).

Table 2
General AMSTAR System Description, Space Segment

JL Transponder
- High-power linear transponder
- 120-W PEP output
- Mode JL: 2 meters and 24 cm up; 70 cm down
- 500-kHz-bandwidth downlink (approx 175 kHz at 2 m; 325 kHz at 24 cm)
- Global beam coverage, all bands
- Spacecraft antenna gain:
  2 m: 2.1 dBi
  70 cm: 12.5 dBi
  24 cm: 16.0 dBi

S Transponder
- Medium-power linear transponder
- 50-W PEP output
- Mode S: 24 cm (1260 MHz) up; 13 cm (2401 MHz) down
- Subtransponders:
  S_1: 100-kHz passband for "normal" mode (FDMA) global communications
  S_2: 100-kHz passband for 20 voice repeater gateway interconnects (TDMA)
  S_3: Packet gateway interconnect; nominally 19.3 kbits/s
  S_4: The S_2 transponder used in broadcast mode
  S_V: Integrated Services Digital Network (ISDN) transponder; 500 kbits/s
- Global beam coverage, uplink and downlink bands
- Spacecraft antenna gain:
  24 cm: 16 dBi
  13 cm: 16 dBi

Microwave Experiment
- Possible 10-GHz stable source for link tests and equipment alignment

Table 3
Preliminary User Equipment Requirements

Mode J
Receive antenna: 15.0 dBi (on-axis)
Preamp noise figure: 1.0 dB
Feed line + misc losses: 1.3 dB
System G/T: −9.8 dB/K
Transmitter power output: 10 W (avg)
Transmit antenna gain: 13.0 dBi (on-axis)
Feed line + misc loss: 1.3 dB
Transmit EIRP: 20.2 dBW (avg) (105 W)

Mode L
Receive antenna: 15.0 dBi (on-axis)
Preamp noise figure: 1.0 dB
Feed line + misc loss: 1.3 dB
System G/T: −9.8 dB/K
Transmitter power output: 10 watts (avg)
Transmit antenna gain: 19.5 dBi (on-axis)
Feedline + misc loss: 1.3 dB
Transmit EIRP: 29.2 dBW (avg) (832 W)

Mode S_1 (General Linear Communications Transponder)
Single dish antenna for TX/RX: 1.5 m (5 feet); dual feed with 50% efficiency.
Receive antenna gain: 28.5 dBi
LNA noise figure: 1.0 dB
Pointing loss: 1.0 dB
Feed line + misc loss: 1.1 dB
System G/T: +4.7 dB/K
Transmit antenna gain: 23.0 dBi
Transmit power output: 10 W (avg)
Transmit misc losses: 1.3 dB
Transmit EIRP: 30.0 dBW (1000 W)

Mode S_2 (Voice Gateway Interconnect)
Same as S_1 station equipment except:
Feed line + misc receive loss: 0.6 dB
Receive noise figure: 0.7 dB
System G/T: +6.1 dB/K

Mode S_3 (Packet Gateway Interconnect: 19.2 kbits/s)
Same as S_2 station equipment

Mode S_4 (Receive Only Gateway Interconnect-Broadcast Mode)
Same as S_2 station equipment
(Mode S_V and microwave beacon user equipment continue under study at this writing)

Table 4
Link Performance

Mode	Avg Downlink S/N	Peak Downlink S/N	E_b/N_0
J	10.5 dB	21.5 dB	12.0 dB
L	11.3 dB	22.3 dB	12.8 dB
S_1	13.4 dB	24.4 dB	14.9 dB
S_2	15.0 dB	33.0 dB[1]	16.5 dB
S_3	—	—	13.2 dB
S_4	21.4 dB	39.4 dB[1]	12.3 dB[2]
S_V	—	—	12.0 dB[3]

[1]ACSSB use assumed; subjective improvement over unprocessed SSB equal to +8dB.
[2]Result obtained if the S_4 Mode were to be used as a dedicated packet link at 32 kbits/s.
[3]At a data rate 500 kbits/s.
[4]The ratio of energy per bit to the reference noise.

S/N ratio on SSB. Peak S/N (the best measure of signal quality in the short term) would be a very respectable 21.5 dB and 22.3 dB for the J and L links, respectively.

Mode S Transponder

Mode S will also fly on Phase 3C next autumn, but it will be a 70-cm to 13-cm version of Mode S and have only limited bandwidth (25 kHz and power of 1.3 W). On Phase 4, however, Mode S will comprise a special 24-cm-up and 13-cm-down transponder, and will provide some truly stunning performance of the transponder. The Phase 4 Mode S transponder is envisioned to comprise four subtransponders, each with its own AGC loops and function. Let's look at the function and performance of each of these subtransponders in more detail (refer to Fig 3).

S_1: General Linear Communications Transponder

The S_1 subtransponder will be used for the traditional type of OSCAR communications most users are currently accustomed to. Essentially, there will be 100 kHz of linear transponder passband for antenna with a dual 24-cm/13-cm feed. A 10-watt average uplink transmitter would produce 1000-watts EIRP using the recommended 23-dB dish gain at 24 cm.

S_2: Gateway Interconnect

The S_2 subtransponder will potentially provide one of the most important services as well as one of the most dramatic. S_2 will be a gateway interconnect transponder. A gateway is simply a portal from one type of network to another. A terrestrial voice repeater can be viewed as a network—a network of users with radios clustered around and interconnected through the repeater. Similarly, the satellite users can be viewed as a network. Interconnection of these networks is accomplished through a gateway. In this context a gateway could be a repeater equipped with an interface to the satellite. That is an uplink transmitter, a downlink receiver and associated interface and control circuitry. Functionally, the gateway serves to extend the repeater user's telecommunication into the satellite's network of users, and vice versa. Ideally, the interface would be transparent; that is, a user in either domain (terrestrial repeater user community or satellite user community) could be totally unaware of the existence of the facilitating gateway. Furthermore, by extension, a terrestrial repeater user linked to the satellite through a gateway could then be further linked through the satellite to a second gateway and its respective user community. Again, if the links were executed properly, users on either end of the dual gateway circuit could be unaware of the extended circuit supporting their QSO.

But there is much more to this gateway arrangement than novelty. Sure, it's amusing to visualize a pair of 2-meter hand-held-radio users half a globe apart enjoying a pleasant chat, describing the radically different scenes before them. But because of the very disposition of equipment within the gateway arrangement, gateway operations using combinations of terrestrial repeaters linked via satellite offer an

the normal Frequency Division Multiple Access (FDMA) use OSCAR users have been employing since AO-6 days. With 100 kHz, there's ample room for about 25 to 35 QSOs, depending on how well they are "packed" or "stacked." If there are three or four individuals per QSO, as there often are in satellite QSOs, about 100 simultaneous users could be accommodated in this S_1 transponder. S_1 performance would average about as good as AO-10 got at its best: S/N of about 13.4 dB. Moreover, under ideal conditions, S_1 could deliver 24.4-dB S/N, peak (see Table 4). In order to realize the specified user S/N, the Mode S_1 user equipment suite (or better) would be required. As seen in Table 3, it consists of a 1.5-m (5-ft) parabolic dish

extremely important approach to emergency communications.

A portable gateway established at a major flood or earthquake site could, for example, link the disaster reaction team to major relief organizations. Support and logistics control could be organized on an unprecedented level. On-scene leaders could communicate instantly with virtually any other QTH in the hemisphere 24 hours a day. A single hand-held radio hiked to a mountaintop airline crash site could communicate directly with state or federal authorities using a gateway on a nearby mountaintop. Establishment of DX communications for local or regional emergency centers could be as simple as implementing the gateway to the continuous coverage satellite(s).

Aside from the unprecedented potential for saving lives and property, gateway facilities would be available for more mundane use between selected repeaters on a daily basis. A limited number of repeater gateways would be authorized access for these routine QSOs when there were no emergency operations underway or if adequate spectrum sharing schemes were to be established. So one age-old fantasy many hams have harbored of having even foot-mobile) while engaging in a DX QSO would be realized simultaneously with the penultimate emergency communications resource!

Moreover, because the real communications "work" involved in communicating the 71,400 km (44,400 mi) or so to/from the geosynchronous satellite is accomplished by the gateway, the equipment burden on the gateway user is reduced to absolute minimum—essentially, only what is needed to communicate over the distance to and from the local gateway/repeater. And that could even be done in some cases with one of those new, ultraminiature 100-mW hand-held rigs now on the market. For a community of terrestrial repeater users who have an interest in linking their repeater to others across the continent, it makes sense to pool their resources to establish a single gateway for the long-haul to/from the satellite, rather than each individual undertaking the cost. Thus, the gateway users sharing the resource would be, in effect, establishing a Time Division Multiple Access (TDMA) system for communicating with the world outside their local repeater community on a given "channel," one of several FDMA channels available.

Compare this TDMA access to the FDMA access users of the S_1 transponder enjoy. The S_1 FDMA user undertakes his own uplink/downlink burden. It costs him the equipment required to establish the link. For this investment he obtains time-independent use of the S_1 linear transponder, ie, he can use it whenever he cares to. On the other hand, the gateway TDMA user, having pooled the uplink/downlink resource in the form of the gateway equipment, may have to queue up to use the resource, ie, wait until it is free for his use. Thus, he has reduced his personal equipment burden at the cost of time-independent QSOing; he's time-sharing the resource with others.

To establish a gateway QSO, the user could simply pick up his hand-held and tap out a few numbers on the DTMF pad to instruct the terrestrial repeater to enable gateway mode. When the gateway replied with a signal indicating the satellite's Demand Assignment Multiple Access (DAMA) facility had responded, indicating a vacant channel pair was available, the gateway user would then tap out the code for the other gateway repeater he wanted to link to. The DAMA facility would then assign a channel pair to the originating gateway and the target gateway, and the link would be established for a preset time period. Users of the originating repeater would then be in contact with users of the target repeater.

The technology to achieve this type of circuit is not new. It derives straight from the pages of today's terrestrial cellular mobile telephone systems. Amateur Radio implementation of a similar system could be much simpler, however, since much of the redundancy and protection used in cellular mobile radio (to assure privacy and avoid misconnects) could be eliminated. It's obvious the S_2 subtransponder could spur enormous achievements in emergency as well as routine communications.

S_3: Packet Gateway Interconnect

Packet radio is generally acknowledged to be the area in which Amateur Radio is currently experiencing the fastest growth. Nearly 20,000 packeteers are now active, according to some sources. That's probably 10-15% of all active US amateurs. The proportion is expected to grow significantly in the last years of this decade. Local Area Networks (LAN) established around a digipeater hub have been linked to other LANs through VHF, UHF and even HF links. Coast-to-coast connectivity, albeit nonistantaneous, is now a fact. Messages dropped in specific packet-radio nodes often reach an individual destination addressee in a day or less. And they arrive there error-free.

The growth of the terrestrial networks is progressing in a step-wise, part directed, part random pattern. Interconnection between widely separated digipeaters on the East Coast and West Coast and some places in between is now possible. But what if these LANs and groups of LANs could be linked by satellite into a continental or even multicontinental network? That's exactly what the S_3 Packet Gateway Interconnect transponder is about. It could link dozens, even hundreds, of packet gateways together with a high-speed trunk. While our initial calculations were made based on a 19.2-kbits/s data rate, the trunk bandwidth could even be up to 56 kbits/s or more if projected-use estimates indicate more resource is warranted.

Recreational use of the packet gateway transponder would, of course, be part of its mission statement. But there is much more to it than merely the digital ragchew, even the DX digital ragchew. Just as the essential "justifying" rationale for the S_2 voice gateway interconnect transponder is the facility and capacity to provide unprecedented emergency communications capability, so, too, would the packet gateway interconnect transponder open new modes of public service. Today's Amateur Radio communicators are coming to well appreciate the tremendous benefits packet radio has over more traditional modes such as CW and even RTTY. Packet-radio messages are error-free, high-speed and self-documenting. Traffic handling, routing, sorting, etc, can all be automated. The result is often remarkable improvements in traffic throughput, accuracy and, most important, communications effectiveness. Portable packet terminals installed on jeeps, rescue trucks and the like are now appearing in and among forward-thinking Amateur Radio emergency-communications communities.

The S_3 transponder aims to afford the emergency LAN a port to a wider community. As required, the field operations center and even portables could communicate with regional or even national emergency-management centers to communicate status, request specific support and implement actions directed by headquarters via this channel. As with the S_2 voice gateway interconnect, S_3 would be available for recreational use, but earn its keep in providing unique emergency and general public-service communications resources as required.

S_4: Broadcast Mode Gateway Interconnect

S_4 is not a separate transponder, but rather a different mode of employment of the S_2 voice gateway interconnect subtransponder. By reallocating on-board resources, a broadcast capability of notable proportions could be established. As shown in Table 4, nearly 40-dB peak S/N ratio might be obtained using advanced SSB techniques. (Amplitude-compandored single sideband, ACSSB, is one means of achieving this very high level of S/N ratio performance.) That's as good as, and in some cases better than, commercial telephone circuits.

The S_4 Mode might be used for many routine and public-service activities. In routine use, ARRL W1AW bulletins might be sent to groups of terrestrial gateway repeaters. Listeners would use their VHF or UHF hand-held radios to tune in the bulletins on their local repeater. Groups of repeaters could be addressed selectively, say by time zone, by tone-encoded addressing. When a given repeater heard its address on the S_4 Mode downlink, it would interconnect the gateway's downlink receiver to the repeater transmitter to retransmit the audio to the repeater's coverage area. Local repeater operators could, of course, override the linking signal at will with local, manual intervention.

However, in the event of an emergency, groups of repeater gateways could be called up using the tone-activated alert scheme. In this way, news of regional or more general emergencies could be flashed to hundreds, even thousands, of repeaters in a few seconds. Imagine the improvements in emergency response afforded. When combined with existing emergency communications structures at the regional and state level, the result could be unprecedented effectiveness in response to earthquakes, hurricanes, general tornado activity, sudden flood emergencies, and so forth.

On the more routine side again, the S_4 mode could help unify Amateur Radio by facilitating the teleconference radio net concept, which to this point has relied on terrestrial telephone network linking of a hundred or more repeaters several times per year. Imagine this concept expanded to several thousand repeaters on line. Moreover, the equipment requirements for a S_4 Mode Receive Only (RO) gateway are quite moderate. As shown in Table 3, a 1.5-m dish with a single 2.4-GHz feed, a routine LNA and a mixer to a convenient IF are all that would be required. By the time the S_4 mode flies, one could likely establish an S_4 RO gateway facility for $300 or less!

S_V: The Mode S Video Subtransponder

Advances in digital television and video data-compression techniques suggest to us there may at last be a good mesh between amateur TV (ATV) and OSCAR satellites. Previously, constraints of power and bandwidth have made anything but occasional forays with slow-scan TV (SSTV) impossible on OSCAR. Now, however, using video data compression techniques we believe it possible to include a transponder capable of relaying digital video at the rate of perhaps 500 kbits/s. Commercial and military developments using comparable rates are very encouraging. Thus, we have every reason to believe these leading-edge techniques will be available to advanced amateurs by the time S_V is on line.

A more general view of the S_V transponder is that it is a general-purpose, high-speed transponder and that it could (should) be configured to handle the Amateur Radio equivalent of the Integrated Services Digital Network (ISDN) now being fielded by telecommunications companies throughout the world. If this were done, bulk file transfer could be accomplished at astounding rates. The types of services that could be provided with the S_V transponder beyond these examples are numerous. Distribution of Amateur Radio software, articles and research papers are some examples that come to mind.

Using the S_V transponder as an ISDN facility for digital video, very-high-speed packet, digitized voice, file transfer, some combination of these or some new, presently unforecast mode is a matter for our study teams and the Amateur Radio community to decide. But it seems clear that this area could be as fertile as our collective imagination.

Microwave Experiment

A further module that could be included on board Phase 4 is a microwave-beacon experiment. Much work is being done using narrowband emissions as high as X-band (10 GHz). Imagine having a permanent 10-GHz beacon aboard AMSTAR to align antenna feeds, tweak LNAs and calibrate antenna positioning equipment. Such a field alignment tool might go far in advancing both interest and proficiency in the SHF bands. This experiment continues under study for possible inclusion.

Conclusion

Traditional OSCAR users have been a specialized lot. They have enjoyed many of the occasionally esoteric challenges becoming highly proficient on OSCAR involves. Tracking and figuring access are not bothersome chores but rather part of the fun to this dedicated bunch. But clearly the view of what's fun and what's not depends on one's interest. Certainly, an emergency communicator is less interested in calculating access to a satellite than communicating his emergency traffic! So unless something changes, OSCAR use will remain a special art practiced by a relatively small group of aficionados.

But it is now abundantly clear that the nature of the satellite game is about to change dramatically with the advent of Phase 4. These changes come about from two fundamental causes:

1) Maturation of OSCAR technology and technologists to where the media becomes transparent to the user, whereas previously the medium was in large measure part of the message (or reason for being on OSCAR). Thus, rather than evolve to further refinements of a traditional theme, OSCAR will be revolutionized to become a utility available to virtually anyone who wishes to participate. Acquisition of special equipment and skills will be minimized and, in essence, consolidated in the gateway concept. There, many participants can share the cost burden. The esoteric aspects of satellite communication can be offset and eliminated by more sophisticated engineering than has ever been incorporated. In sum, it is the highest form of the engineering discipline to make the inherently complex seem simple and generally accessible.

2) There is a growing, urgent need to make productive use of our incalculably valuable spectral resources. Where commercial interests see our UHF spectrum quite literally in terms of gigabucks (billions), you must be convinced of the pressure to abscond with the heart of our hobby (our frequencies) will become enormous. We simply *must* do better to justify our continued occupancy of the UHF bands, lest we lose them forever. Far from being the sounds of distant cannons, the threat is clear and present. If we don't move now, we could very well face significant challenges for our spectrum at the next World Administrative Radio Conference (WARC)—or even sooner if the FCC opts to change those secondary allocations. An Amateur Radio satellite using key UHF frequencies in providing real, tangible, demonstrable public service on a regular basis is one of the best ways we know to ensure we retain our spectral resources. Building Phase 4 and using it for the general public benefit is not just a further expression of altruism, then, but an element in the preservation of our most valuable resource–spectrum–for decades to come. We *must* make better use of our UHF spectrum soon or it surely *will* be gone!

The challenge of Phase 4 is this: Come to understand the potential for unprecedented levels of public service and technical achievement; develop the plan to implement the system that manifests the potential and wisely manage the powerful resource that results.

Is Amateur Radio up to this challenge? We obviously believe so, or we would not have brought this preliminary vignette to your attention. We sincerely believe Phase 4 will be operational in about five years and that it will forever change the nature of our hobby. To realize its full potential, however, substantial effort must be dedicated to first eliciting suggestions on meshing the strawman system with actual needs of the user community. For example, the operational requirements of the emergency communications community are best known by the emergency communicators. The direction and objectives of the packet-radio activity are best known to the packeteers, etc. Thus, one of AMSAT's main challenges is to "network" (establish working relations with) its system architects and engineers with the user communities.

To that end, AMSAT is briefing leaders in various Amateur Radio communities regarding the nature of the project and progress toward specific goals. Conversely, AMSAT is actively seeking inputs on technical and organizational matters. Would-be participants should understand at the outset, however, that this is a long-term project that will require comparably long-term dedication by the participants. ATVers, microwave experimenters, repeater organizations, emergency communicators, traffic networks, packet-radio users and all those with a long-term interest are invited to share their ideas on Phase 4 and potential applications. Invitations to participate in applications research studies will be issued in 1987 to individuals and groups who may contribute to the program. Expressions of interest may be sent to AMSAT, Phase 4 Program Manager, PO Box 27, Washington, DC 20044. (Please include a business-size SASE if a reply is sought.)[1]

Phase 4 can change Amateur Radio for the better by providing real public service while simultaneously providing space-age telecommunications to a broad cross-section of Amateur Radio. In that sense, it's not something that we would *like* to do, but rather something we simply *must* do!

---

[1]AMSAT membership is open to the public. Members receive the biweekly newsletter, *Amateur Satellite Report*, and other benefits. Inquire about membership and how to get started in OSCAR by writing to AMSAT.

A Mode-L Parabolic Antenna and Feedhorn for OSCAR 10

Get away from the crowded HF bands. Open new doors with this easily built Mode-L dish.

By Eugene F. Ruperto, W3KH
RD 1 Box 366
West Alexander, PA 15376

OSCAR 10 is the space-age answer to poor or crowded conditions encountered on the lower amateur bands. Sitting here at or near the bottom of the sunspot cycle has really put a crimp in my DXing habits lately, so I decided to put some effort into my OSCAR station. It seems like only yesterday that I listened to the weak CW "HI-HI" from one of the earlier OSCARs and marveled at the technology behind the effort. Later I taped and compared the analog audio telemetry from another OSCAR and wondered where it would end. Throughout the years, I have worked through all of the amateur satellites and mourned the death of AO-7 where I cut my teeth on Mode B.

Now I can sit at the OSCAR station for hours and work at my leisure. No more hurry-up, 10-minute passes. No more frantic antenna positioning. OSCAR 10, with its high elliptical orbit, is "a set it and forget it" type of satellite, for all practical purposes. Being a chronic AO-10 user, I have a tendency to become complacent and take AO-10 for granted. I have access to it nearly every day of the week, and have my choice of ragchewing, working DX, slow- or high-speed CW, certificate chasing, RTTY or just some old-fashioned antenna experimenting. This "bird" gets my vote for being the best so far. In mid-1986, another Phase III satellite with Mode-B and Mode-L transponders will be placed in orbit.

Each satellite poses a new challenge for users and contributes something new to the amateur space program as we know it today. Satellite experiments involving preamplifiers, antenna configurations, polarity or power changes provide instantaneous feedback for analysis. In this respect, AO-10 represents a test bed for the amateur. Now is the time to get the bugs out of your present satellite system and, if possible, explore the world of Mode L. For future launches, the trend is toward UHF, and the amateur with only Mode-A or -B equipment will not be able to take advantage of all that is available.

Since discovering the Mode-L beacon, I've been interested in getting something going. I monitored the frequency when the bird switched from Mode B and found stations on Mode L that are using modest power levels and simple antennas on the downlink. This gave me the incentive to design an uplink antenna for Mode L. Despite the fact that the Mode-L transponder was not performing as planned, what I heard and read confirmed that with a modest addition to the Mode-B station, Mode L could be used. With a 2-m/23-cm transverter at hand, only a small amplifier and a suitable antenna were needed to use Mode L.

Design Strategy

I decided that because of the small antenna sizes required at this frequency (1269.5 MHz) a parabolic dish antenna would be best. Some of the stations on Mode L are using commercially built TV dishes. But, considering the work needed to put them together, cover them with screen, build a feed antenna and supports, add a mount, and live with a deep f/D (usually 0.375), the buyer basically gets only a preformed wire grid for his money. I decided to build my own dish from readily available materials. This custom-design approach offers the builder some measure of control over the finished product. By using this method, one can design and construct parabolic reflectors of any reasonable size for any phase of amateur endeavor.

I used a three-step approach to the design. First, the dish should be effective, which means that a study of size vs power must be considered. My calculations indicated that a gain between 21 dBi and 27 dBi would be sufficient to hear my signal through the 436-MHz downlink, provided

Table 1
Design Considerations for Parabolic Dish Antennas

Dish Size	4†		5†		6†	
Beamwidth (Deg)	13.2		10.5		8.7	
Gain (dBi)	21.8		23.7		25.3	
Output Power (W)	EIRP		EIRP		EIRP	
100 W	16.2 kW		25.4 kW		36 kW	
50 W	8 kW		12.75 kW		18.3 kW	
30 W	4.8 kW		7.65 kW		11 kW	
15 W	2.4 kW		3.75 kW		5.4 kW	
10 W	1.6 kW		2.5 kW		3.6 kW	
5 W	800 W		1.25 kW		1.8 kW	
2.5 W	400 W		625 W		900 W	

†Based upon 50% reflection efficiency and 10-dB taper feedhorn for 0.4 f/D dish

Table 2
Detail Design for Three Parabolic Dish Antenna Sizes

Dish Size (Feet)	Y, X Coord (In)*	Focal Length (In)	Depth of Dish (In)	Gain (dBi)	Beamwidth (Degrees)
4	0, 0	19.2	7.5	21.8	13.8
	3 0.117				
	6 0.468				
	9 1.054				
	12 1.875				
	15 2.929				
	18 4.218				
	21 5.742				
	24 7.5				
5	0, 0	24	9.375	23.76	10.54
	3 0.093				
	6 0.375				
	9 0.843				
	12 1.5				
	15 2.343				
	18 3.375				
	21 4.594				
	24 6.0				
	27 7.60				
	30 9.375				
6	0, 0	28.8	11.25	25.35	8.78
	3 0.078				
	6 0.312				
	9 0.703				
	12 1.25				
	15 1.953				
	18 2.812				
	21 3.828				
	24 5.0				
	27 6.328				
	30 7.812				
	33 9.453				
	36 11.250				

*Cartesian coordinates derived from $y^2 = 4FX$.
y = radius of dish, x = axis of focus, 0, 0 = dish center.

that the combination of RF output power and dish gain would allow me to reach the target value of 3 kW of EIRP (see Table 1). Second, the dish should be light so that it can be mounted at the mast with my other satellite antennas. This prevents an aiming discrepancy later on (such as my downlink antenna pointing at the bird and my uplink antenna pointing toward some meaningless position in space). This becomes more apparent as dish size increases, which brings up the third requirement—beamwidth. A wide-beamwidth dish is nice to have for a moving satellite, but it means a sacrifice in gain. On the other hand, too large a dish will narrow the beamwidth sufficiently to make tracking a chore. Although the Phase III satellites don't require much beam pointing near apogee, the Mode-L operating period is sometimes changed to a lower mean anomaly for operational reasons, and the ground track can cover a considerable angle in terms of antenna movement. It's a nice feeling to not have to aim the antenna more than once every 10 or 15 minutes during the Mode-L period. I decided that a 5-foot dish, with 12 watts of power at the feedhorn, should do the job for me.

Materials

I used ¼-inch-diameter steel pencil rod (a mild steel) for the rib construction. With a weight of 0.167 pounds per foot, it would allow the antenna to be light enough to rotate with the other antennas. This pencil stock is usually sold in 20-foot lengths, called "joints," and at this length they are extremely flexible. To get them home, plan to cut them in half with a hacksaw or carry them, as I did, on an extension ladder on top of the pickup. Keeping the original length will minimize waste when making your cuts. At this writing, the cost of a 20-foot joint was $1.30. Roughly figured, five joints will be needed for a 5-foot dish. Excluding the price of a piece of pipe and three hose clamps, the steel needed for the 5-foot dish will set you back about $6.50, which is a lot less than the cost of your average catalog TV dish. Previous experience using ¼-inch hardware cloth and wire ties for dish coverings led me to abandon that approach in favor of a suggestion by Richard Dolenc, WB3CRF, to use common aluminum screen-door material instead. An added bonus is the lack of bandages required for the inevitable cuts that result from using hardware cloth. The door screen material mesh size is much smaller than the one-tenth wavelength required to be a perfect reflector at this frequency, but it also presents greater wind loading, with the probability of ice and snow accumulation. The fact that the dish can be pointed to a "stored" position when not in use, however, minimizes these effects. A 16-foot-long by 3-foot-wide piece of aluminum screen, obtained at the local hardware store for $6, was more than ample for the 5-foot dish.

A very neat, taut covering can be accomplished by using Liquid Nail™ or similar material to glue the screen to the ribs. Use a caulking tube (price less than $2) to apply the glue. Make sure the glue is waterproof since it will be exposed to the weather. A question arises concerning possible galvanic action between the steel rod and the aluminum screen. The glue matrix acts as a buffer between the metals and, for the most part, as an insulator between the two. As a consequence, only very small areas will occasionally make contact. Though these could disintegrate over a period of time, the glued surface area will be much larger and still retain the original strength. The entire screen surface may be removed with a pair of scissors or a sharp knife; after grinding the rib surfaces free of glue, a new screen surface can be applied.

Dish Construction

A ¼-inch-thick piece of scrap steel plate, 2 feet wide by 4 feet long, is used as a jig to build the ribs. The parabolic X and Y coordinates (see Table 2) are transferred to the plate, spaced roughly every three inches, by tracing the figure on the plate. Short pieces of ¼-inch steel strap are tack welded along the curve to hold the pencil rod along the front of the curve; then similar pieces are tack welded along the back to hold the pencil rod at the desired shape (see Figs 1 and 2). The process is repeated for the back rib member as well as the piece of pencil rod that forms the hub portion of the rib. The hub pieces extend about 2 inches beyond the rib dimensions so that when completed, a hose clamp can be fitted on both front and back of the rib to clamp around a slightly longer piece of ½-inch pipe, threaded at both ends, that forms the central hub.

Once the jig is completed, and the first rib welded and checked, the remaining five ribs can be constructed in a short time. The

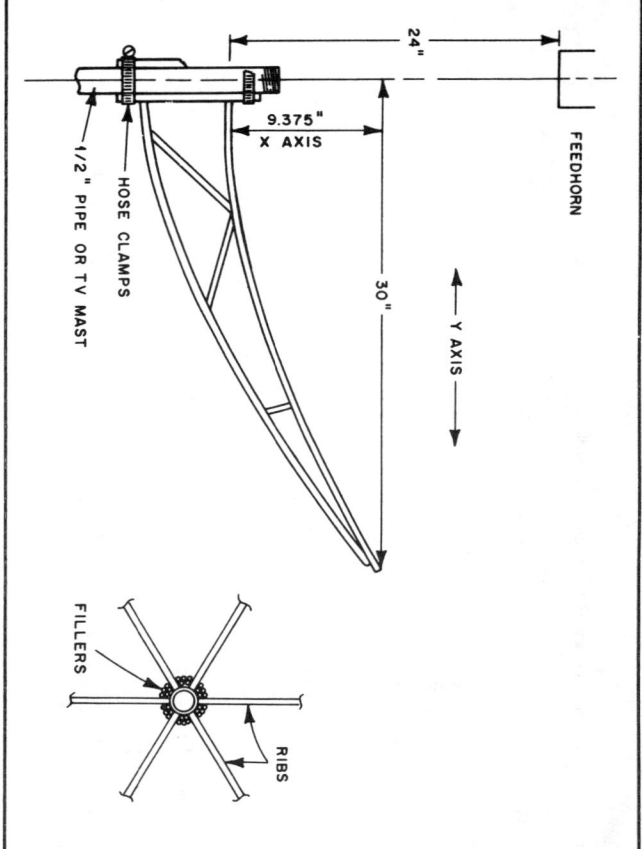

Fig 1—Rib construction details. Dimensions are as specified in Table 2.

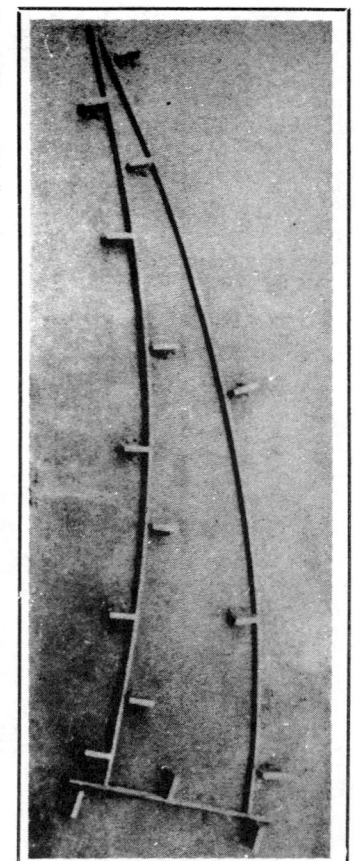

Fig 2—Rib assembly jig.

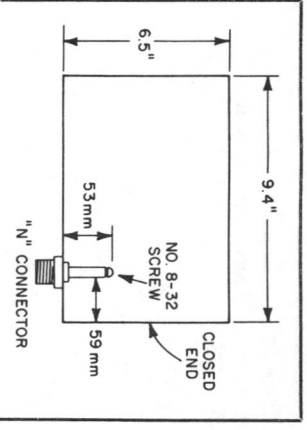

Fig 3—Feedhorn construction dimensions. See text for details.

rib accuracy is checked using a plywood curve gauge cut out on a bandsaw. Because of the 6-rib design, a sloppy fit results when the ribs are clamped to the hub piece. Insert additional 9-inch pieces of pencil rod as fillers to make a tighter fit when attaching the ribs to the ½-inch pipe hub. Place the ribs and pipe "face down" to space the ribs evenly. I used a joint of pencil rod, bent to a diameter of 5 feet around a tractor tire, and welded it to the outer tip of each rib before tightening the hose clamps. Alternatively, single pieces of rod may be cut to fit between the ribs at the edge of the dish to form the rim. The completed dish, with center pipe hub, weighs approximately 12 pounds. More ribs could be added, at the expense of additional weight.

Cut sectors of aluminum screen that are slightly larger than the rib spacing so that the sectors will overlap. Run a bead of Liquid Nail, or other glue, on the front side of two ribs and the included part of the external ring and press the screen sector in place. The glue dries fairly rapidly at room temperature. Run a strip of thin wood over the screen surface to tighten it and allow the glue to penetrate and hold the screen surface. By the time this operation is complete, the next sector can be applied in the same manner.

Feedhorn Construction

There is a plethora of tin-can feedhorn designs described in other articles. I prefer designs that use a full guide wavelength in their construction. At the risk of drawing flack from the experts, I believe that the feed acts better if designed at half the guide wavelength, and it reduces the amount of tin in the construction and the weight by half. H. J. Griem, DJ1SL, reported experiments on a tubular horn for the 13-cm band.[1] He discovered that by using a full guide wavelength for the horn, a resonance lower in frequency exists outside the amateur band that corresponds to one-half of the guide length, which is, of course, one full guide wavelength at 13 cm. This effectively narrows the bandwidth of the 13-cm horn, making probe tuning more critical. By using half of the calculated guide length, the bandwidth is increased and probe tuning is less critical. Using this method, several successful feeds have been constructed for a variety of microwave frequencies. The design described here has an SWR of 1.15:1 at 1269.5 MHz, measured at the feedhorn.

The feedhorn is mounted on a small plywood saddle attached to a length of ¼-inch pipe by two small U clamps. The ¼-inch pipe is offset at the center of the dish with a ¾-inch, 90° elbow, a short nipple, another 90° elbow, another short nipple, and a reducer from ¾ to ½ inch (see Fig 4). I prefer a tripod method of mounting to this method because the ¼-inch pipe passes close to the mouth of the horn.

Feedhorns can be easily replaced, so we can try a switchable left- or right-hand circularly polarized feed at a later date. It might be worthwhile to mention that most feedhorn designs use a full guide wavelength in their construction. At the risk of drawing flack from the experts, polarization complicates construction, especially when you're in a hurry. Feedhorns can be easily replaced, so we can try a switchable left- or right-hand circularly polarized feed at a later date. It might be worthwhile to mention that most feedhorn designs use a full guide wavelength in their construction. At the risk of drawing flack from the experts, I use a linear-polarized feed for several reasons. Most of the stations heard on Mode L are using linear polarization, and the downlink signals sound okay. Circular polarization complicates construction, especially when you're in a hurry. Feedhorns can be easily replaced, so we can try a switchable left- or right-hand circularly polarized feed at a later date. It might be worthwhile to mention that most stations heard on Mode L are using linear polarization, and the downlink signals sound okay. Circular polarization complicates construction, especially when you're in a hurry.

I use a linear-polarized feed for several reasons. Most of the stations heard on Mode L are using linear polarization, and the downlink signals sound okay. Circular polarization complicates construction, especially when you're in a hurry. Feedhorns can be easily replaced, so we can try a switchable left- or right-hand circularly polarized feed at a later date. It might be worthwhile to mention that most feedhorns can be easily replaced, so we can try a switchable left- or right-hand circularly polarized feed at a later date. It might be worthwhile to mention that most feedhorns can be easily replaced.

The feedhorn used here was constructed to my specifications by Dick Dolenc, WB3CRF, using approximately 26-gauge galvanized steel capped and soldered at one end. It has only one seam running longitudinally along the feed axis. A piece of copper tubing, threaded to accommodate a no. 8-32 bolt for tuning, is soldered to an "N" connector (see Fig 3). The "N" connector is then soldered with a propane torch into a 5/8-inch hole, located as shown in Fig 3. Apply two coats of paint to the outside of the horn for protection.

[1]H. J. Griem, DJ1SL, "Tubular Radiator for Parabolic Antennas on the 13-cm Band," *VHF Communications*, Apr 1976.

Conclusion

The rib design is very strong. During initial testing, the dish took an unscheduled trip across the backyard during a severe windstorm. It suffered only one small deformation on the outer ring and a bent feedhorn mount, both of which were easily repaired. I surprised myself by listening to my signals on the Mode-L downlink the first time. Running only 12 W at the feed and pointing the dish approximately at the bird, I worked almost everybody I heard.

The antenna is now tower mounted, and I have increased the power to 40 W at the feedhorn. Performance is very satisfactory, and the pointing angles for tracking are reasonable. I found out another thing, though—you need a better receiving system on Mode L than on Mode B. This may be because the Mode-L transponder is not up to performance standards because of component failure.

Build a dish and try Mode L. It sure is lonesome up there now, but that will change!

Fig 4—Completed dish assembly. Note the feedhorn mounting scheme.

Microcomputer Processing of UoSat-OSCAR 9 Telemetry

Are you interested in what satellites are "saying"? Here are some pointers to get you started examining satellite-transmitted data.

By Robert J. Diersing,* N5AHD

UoSAT-OSCAR 9 was built by members of the Electrical Engineering department of the University of Surrey, England. The satellite was placed into orbit on October 6, 1981. An on-board telemetry system provides data derived from monitoring 60 analog sensor channels and 45 digital status points. Analyzing the data can be a fascinating pastime. (A detailed description of UoSAT-OSCAR 9 may be found in *The Satellite Experimenter's Handbook*, published by the ARRL.)

A second UoSAT, OSCAR 11, was launched on March 1 of this year. The satellite was initially silent, but the engineers and scientists have restored it to

*Assistant Professor of Computer Science, Corpus Christi State University, 4129 Montego, Corpus Christi, TX 78411

operation. It is now transmitting telemetry while its condition is evaluated.

Satellite Telemetry System

UoSAT-OSCAR 9 transmits the systems status and experiment measurements in ASCII using FSK with 1200- and 2400-Hz tones and even parity. These frequencies are close enough to the Bell 202 standard tones of 1200- and 2200-Hz that a type 202 modem will work well. (UoSAT-OSCAR 11 tone frequencies are reversed from the Bell 202 standard in their binary meaning.) The data rate can vary between 110 and 1200 bauds, but 1200 bauds is the rate most used.

Different telemetry formats are in use. These are shown in Figs. 1-4. The format shown in Fig. 1 is the older, standard form, combining the spacecraft status and telemetry values. Of the two newer formats, that shown in Fig. 2 has the same 60 telemetry values, but with the spacecraft status deleted and a checksum added for each value. The Fig. 3 format is one in which only certain channels are transmitted repeatedly after having been recorded at regular time intervals during the entire orbit. A sample of the UoSAT-OSCAR 11 telemetry is shown in Fig. 4. I'll concentrate on describing how to get the telemetry data into a computer in a form that will allow you to analyze it within the limitations of your hardware and programming experience.

Telemetry Reception and Capture System

The system in use at N5AHD consists of several processes: (1) orbit prediction, to know when to listen; (2) data capture, live or on audio tape; (3) demodulation of the data and its storage on diskette; (4) editing of the raw data to exclude detectable errors; and (5) analysis and display

```
AMSAT 10101 10000 00000 01110 00011 11001 00000
AMSAT 10101 10000 00000 01110 00011 11001 00000
00110 01001 02765 03001 04001 05433 06370 07303 08486 09482
10100 01160 12000 13366 14314 15188 16420 17885 18442 19438
20170 21470 22724 23024 24006 25422 26419 27267 28493 29611
30280 31180 32666 33235 34012 35333 36401 37401 38509 39313
40070 41110 42736 43102 44044 45000 46002 47467 48526 49502
50070 51000 52274 53089 54637 55450 56463 57488 58486 59507
```

Fig. 1 — Standard format of a UoSAT-OSCAR 9 telemetry frame. The first two lines indicate which spacecraft systems are active. Telemetered values from the spacecraft systems and experiments are contained in the next six lines.

```
UoSat Computer-generated checksummed telemetry
Format: Per channel, sum 5 data digits (0-9), print as (A-Z, a-p)

0011001190L0276R03003010E04001F05620066870768110864425094663W
10100C11080K12000D13370014311K15660S16572V117234R1837T3W194000
20150121160K22727D23117020140H254180Z6428W27283W28458B29584c
30290031430132668Z32567T34011J35366X363996373472T438475b39209X
40090N41090042744V43019R44141045001K46003N47410048497B494446b
50100651090P522745309154930V55412R56458c57458d58428B59463b
```

Fig. 2 — The checksummed standard telemetry format. Status lines are deleted, and the letter between measurements can be used to check the validity of the preceding five digits. The message shown above the frame is transmitted by the spacecraft before each frame.

```
0BD50140064002400140024008003024002407A        01077086680000006680088629
0BDC0540C119D6750440E1870440F449094016         0108708668000000668008628
0BE00240054008400540054003400240800391         0109708668000000668006901F
0BE70540024001409372014002400224024017         010A0866700000006680055450
```

Fig. 3 — These columns show two examples of whole-orbit telemetry dump format. With this form, several channels can be sampled by the spacecraft throughout an entire orbit, and the information retransmitted. The data consists of a frame sequence number followed by the measured values and a checksum. Usually, the weekend code-store will tell when the whole-orbit data was collected and what channels were included from the previous week.

```
UOSAT-2                                        840224522100
005151011039B02011203010204023505028F06025107031508032909026D
1051501100012005613013140051150041600071773641873619736A
2051532103222667723000124001725000726077427736728736829736 9
305152310165322844F3000036000730530536000537736638353E39353F
40763641000542688043000744000044550564600024773614835394934 6C
50561751017252661653261541110558525F56000357360758735593539
60210561708C76280006300416410036510DE6614C056734066800069000F
```

Fig. 4 — UoSAT-OSCAR 11 checksummed telemetry sample. This is the most common format transmitted to date, but other formats are possible during data collection for and after attitude maneuvers.

of the captured data (see Fig. 5).

Software Configuration

The software I use is written in several programming languages for various reasons. The orbital prediction phase is handled by a program written in PL/I-80.[1] I prefer to do the orbit-prediction phase with a program that compiles to machine language rather than BASIC, which is much slower.

Data capture is done with one of two programs, both of which are written in Z80™ assembly language. One program captures the received characters by polling the serial port to which the modem is attached. It places the characters into a buffer, whose contents can later be transferred to disk. The other program uses interrupts to capture the received characters from the serial port and place them into a buffer. In the meantime, data is taken out of the buffer and sent to another computer for real-time display of decoded telemetry.

The data editing and analysis programs used in steps 4 and 5 are also written in PL/I-80. This is done primarily because of the faster execution times and better file-handling features that are available.

Hardware Configuration

I use a Cromemco Z-2D microcomputer. This is an S-100 bus system, and it uses a CP/M™-like operating system called CDOS. I find most programs are transportable between CP/M and CDOS systems; the programs described in this article operate on a CP/M system. The other system components are a Cromemco SCC (single-card computer), a 16FDC floppy-disk controller, 64 kbytes of memory, a TUART (Twin Universal Asynchronous Receiver-Transmitter) digital interface, a Heath H-19 terminal and a Novation 4202B modem.

Data Capture Procedure

Capturing the data transmissions on a quality cassette tape recorder, with the help of a discriminator meter and an audio-level meter, should pose no problems. Even though you may wish to place the data directly into memory, the cassette tape provides an excellent backup in case you run into problems. If you decide to use the computer to capture the data as you receive it, you may have to spend some time reducing computer RFI so your receiver will operate properly.

To capture UO-9 data, the following steps are required:

1) Audio is fed to the modem directly from the receiver or from the audio tape player. When recording, be sure the audio level is not too high. Even though the limiter circuits in commercial modems are good, it would not hurt to pay some attention to impedance matching. You should check to see if the 2400-Hz tone is much lower in level than the 1200-Hz tone; you may have to pick up the audio just after the discriminator rather than at the speaker leads.

2) The modem output is connected to a serial input port of the computer. The physical connections are defined by the RS-232-C standard.

3) Software that will accept the signals from the computer serial port and store the data in memory must be written (or obtained). This software must also be able to save the captured data in a file on an external storage device, such as a floppy disk.

4) The computer must have an external data-storage device (disk drive or cassette tape). This way, the data-analysis programs can process the data without having to make the conversion of analog (audio) signals into digital signals again.

The Data-Capture Problem

The time it takes the computer to process a single character must be less than the time it takes for the next character to arrive. With data arriving at 1200 bauds, it is usually necessary to write the capture program in the computer's native language. This means writing in assembly language for, say, the Z80. Even if the computer's BASIC interpreter allows access to serial ports, BASIC probably will not be fast enough to process data at 1200 bauds. Rather than attempting to teach assembly language programming, I'll show flowcharts for the data-capture program. There will be an explanation of these later.

Serial Ports and Operation

A data "port" can be thought of as a

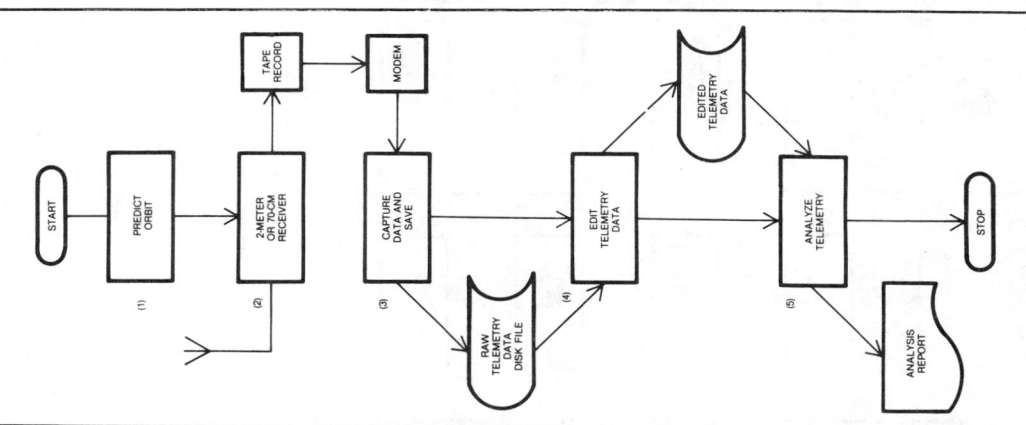

Fig. 5 — Flowchart of the UoSAT-OSCAR 9 telemetry capture and analysis system used at N5AHD.

[1]Orbit-prediction software is available from the AMSAT Software Exchange, Box 27, Washington, DC 20044.

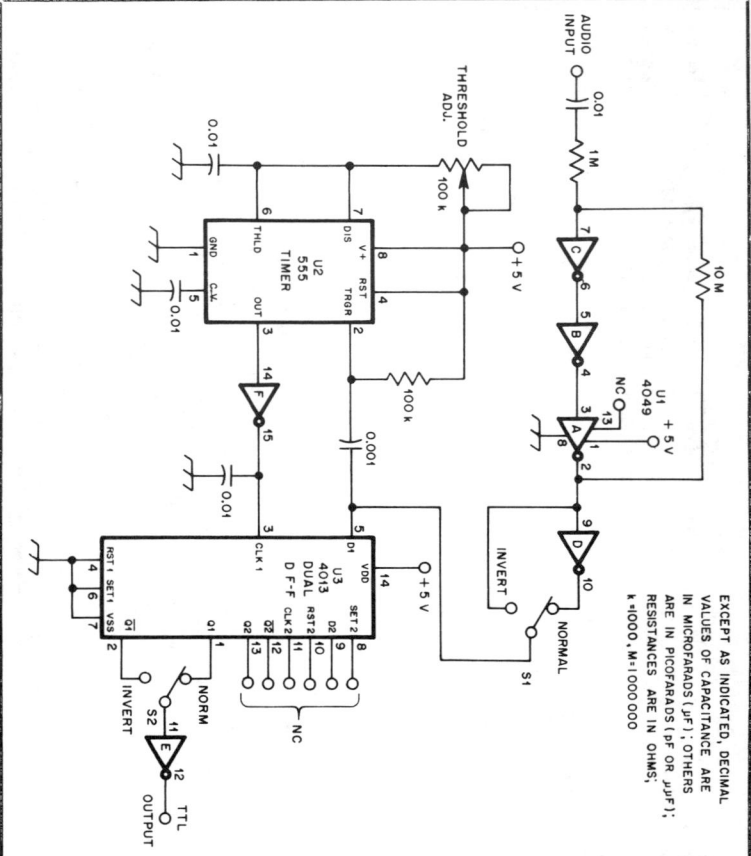

Fig. 6 — A simple demodulator for UoSAT-OSCAR 9 use. The incoming audio should be noise free. This circuit does not regenerate clock pulses, but should work fine for all asynchronous signals. The current drain of this circuit is about 5 mA at 5 V. For initial adjustment, feed an 1800-Hz tone into the input and move the THRESHOLD ADJ. potentiometer until the output of U1 is on the verge of changing state. (tnx to Steve Gomez, KE5O, for this circuit)

mechanism by which the microprocessor has access to the data presented. It is a combination of hardware and software.

Serial transmission and reception is a mode in which one bit at a time is sent or received. Since information is transferred bit by bit, the receiver must know the rate at which the transmitter is sending. In this case, the satellite is the transmitter and the receiver is the computer. If the satellite is transmitting at a rate of 1200 bits per second (bit/s or 1200 bauds), the computer must check for incoming bits at the serial port at a rate of 1200 bauds.

A modem is a *modulator/demodulator*. In this application, the modem changes audio frequency shifts picked up at the radio receiver into different voltage levels to be sent to the computer serial port. The voltage levels should be in accordance with the RS-232-C standard.

Some microcomputers are supplied with serial ports. Check your hardware manuals to see if a serial port is available. You may be able to use the printer port if it is a serial type. If you need to purchase a serial interface, you can generally find them advertised in many microcomputer magazines. Two interfaces I have used are the Cromemco TUART and the Solid State Music IO-4. Both of these have two serial and two parallel ports on a card that plugs into an S-100 bus system. You can also purchase serial interfaces for the Apple® II

About Modems

Where do you get a modem? You have two choices: Build one, or purchase one. The schematic diagram for a simple demodulator is shown in Fig. 6.

If you purchase a modem, be sure it is a Bell type 202 modem and not a 212 type. The type 212 modems are popular for 1200-baud transmission over telephone circuits, but do not operate on the proper tone frequencies for this application, nor do they use FSK at this data rate. Type 202 modems show up from time to time as surplus items,

and Radio Shack TRS-80® microcomputers. Radio Shack model III and IV computers purchased with two disk drives probably already have a serial port. The TRS-80® Color Computer also has a serial printer port.

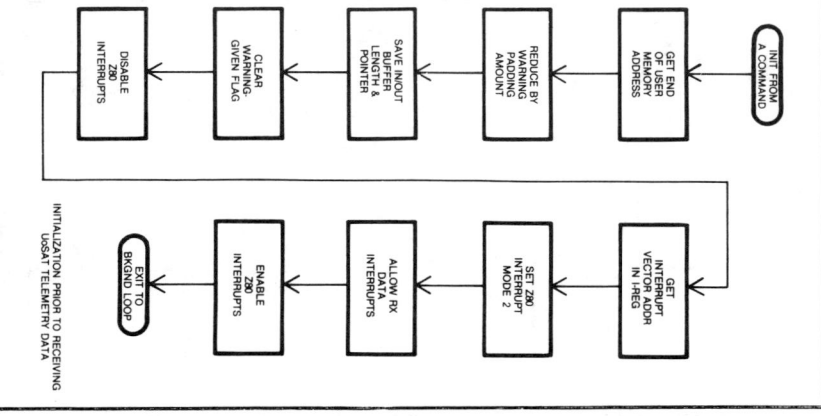

Fig. 7 — Initialization prior to receiving telemetry data.

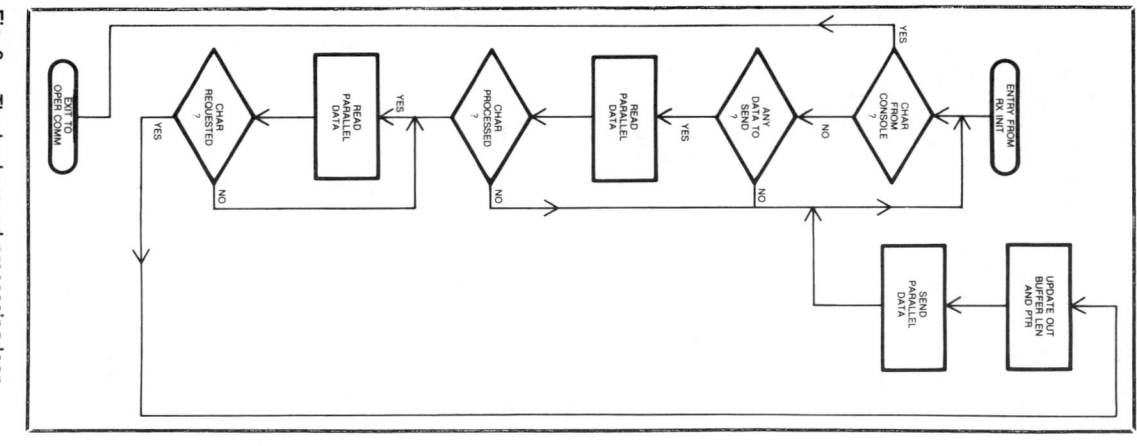

Fig. 8 — The background processing loop communicates with the operator, or transmits data to another computer for decoding and display.

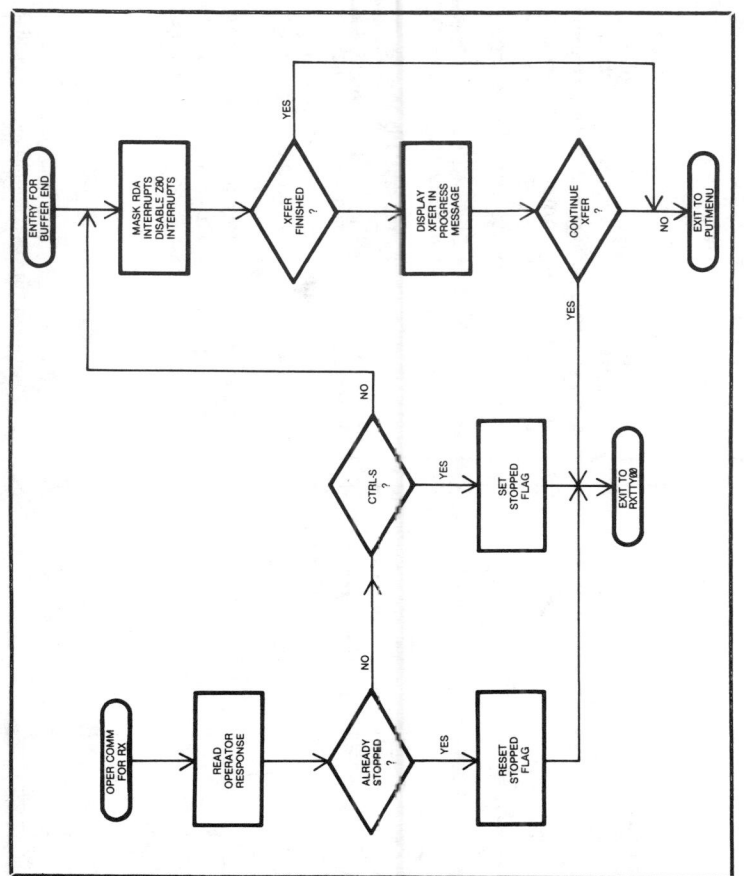

Fig. 9 — Operator communications routine.

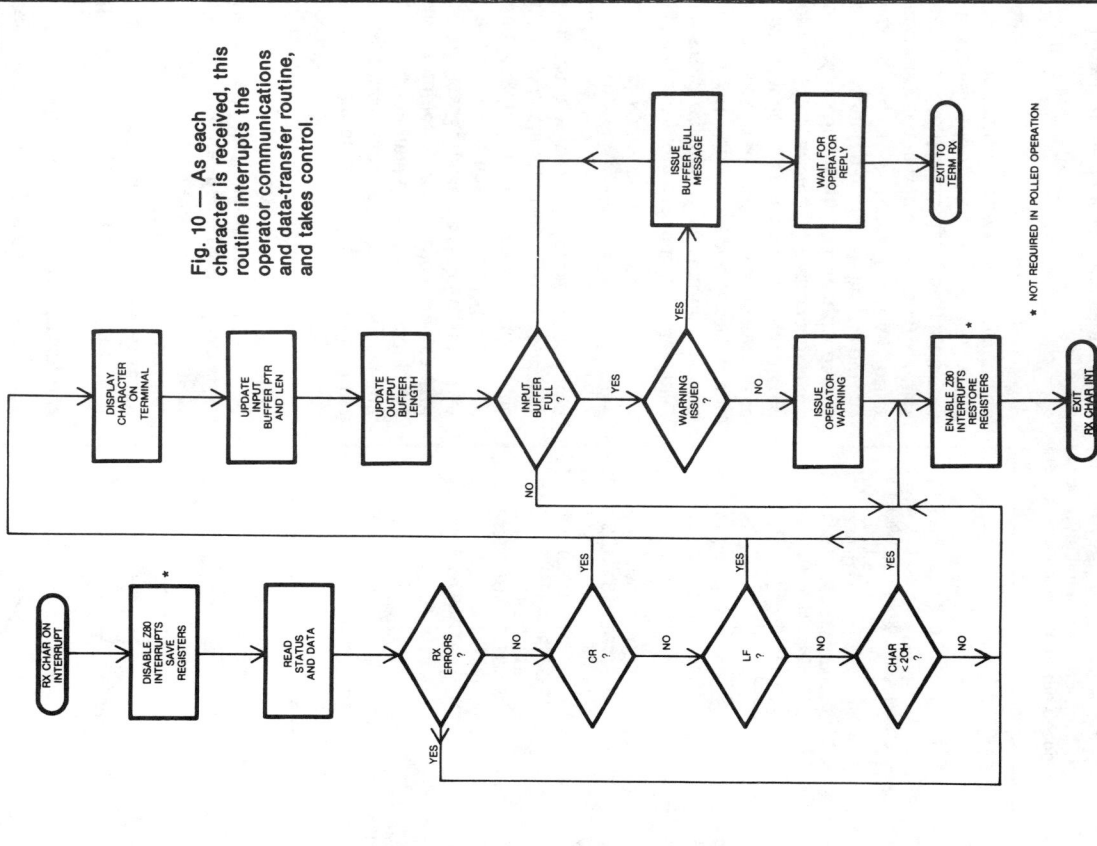

Fig. 10 — As each character is received, this routine interrupts the operator communications and data-transfer routine, and takes control.

* NOT REQUIRED IN POLLED OPERATION

so keep your eyes open for them.

Connecting the Modem and Serial Port

Once the modem and serial interface are on hand, simply connect the serial output of the modem to the serial input of the computer. Only two wires are required: one for data, and one for signal ground. If you have trouble getting data through, the transmit and receive data lines (pins 2 and 3 on the DB-25 connector) might need to be reversed. This is because the RS-232-C specification defines two types of equipment configurations: data terminal equipment (DTE) and data circuit-terminating equipment (DCE). Since these are complementary ends of a circuit, the signals will be reversed at one end. Also, there may be modem signal lines that have to be permanently wired to a logic low or high level. This is because modems control data going in both directions. For our work, the modem needs to be in the receive mode.

All of this may sound complicated, but you will likely find a description of the signals in the modem documentation. Sometimes there are switches inside or on the rear of the modem that change the configuration of some of the signals. If you happen to have a modem that has switches or jumper positions for full-duplex or four-wire operation, you should enable these options.

Software Interface to the Serial Port

Rather than trying to explain the operation of a Z80 (or other) assembly language program, I have divided the functions needed to process serial modem data into separate routines. Flowcharts for these routines are included. Here is a list of the necessary functions and some brief comments about each.

Initialization Prior to Reception

The initialization routine (Fig. 7) must set the operational characteristics (such as the data rate and word length) of the serial port. It is possible that these items are not software programmable, but are hard-wired on the interface. The pointers to the internal received-data buffer must be initialized. If you are detecting received data via interrupts, the proper interrupts must be allowed (unmasked), and the proper interrupt mode for the processor must be specified. It is not necessary for received data to be processed by interrupts. I have included this method because I use it on occasion.

Background Processing

This routine (Fig. 8) executes in between received characters. In a non-interrupt-driven system, it will probably do only two things: check for intervention from the system operator, and see if another

Satellite Anthology 49

character has been received from the modem. In the interrupt-driven system, this routine would still check for operator communications, but the arrival of a new character would be signaled automatically by the interrupt. In my interrupt-driven system, this routine has the additional task of sending the received data to another computer for real-time display.

Operator Communications Routine

At some time during the data-capture process, it may be desirable for the operator to temporarily, or permanently, suspend data capture. The operator communications routine (Fig. 9) processes these requests accordingly. If transmission to another computer is in progress, the operator is warned and can allow it to finish, or abort, the process. If termination is requested and reception is interrupt-driven, the receive-data interrupt must be masked again.

Receive Characters from Modem Routine

If reception is not interrupt-driven, this routine (Fig. 10) would become a part of the operator-communications loop and would be executed if a character is ready to process. In an interrupt-driven system, it is automatically executed as a result of the receive-data interrupt. In either case, the overall function is the same except that in noninterrupt-driven (polled) systems, interrupt-related functions would not be included.

The character-receive routine must accept the character from the modem and perform minimal data error checking. As an example, it could filter out control characters. It must place the character in the buffer and update the buffer pointer and length. Finally, it must decide if the buffer is about to fill up. If so, the operator is given a warning before the condition occurs.

Save Telemetry on Disk

Once reception has ended, the data may be saved on disk. This routine (Fig. 11) must ask the operator for a file name and then check to see if it already exists on the disk. If it does, a new name can be entered or the old file deleted. The amount of space needed is computed, and then the data is moved from the buffer to the I/O buffer, one sector at a time. The only other necessary action is to check for errors after each write to the disk. It is possible that the disk could fill up and the operator would need the chance to save the information on another disk.

Editing and Analysis

Now that the data has been captured on

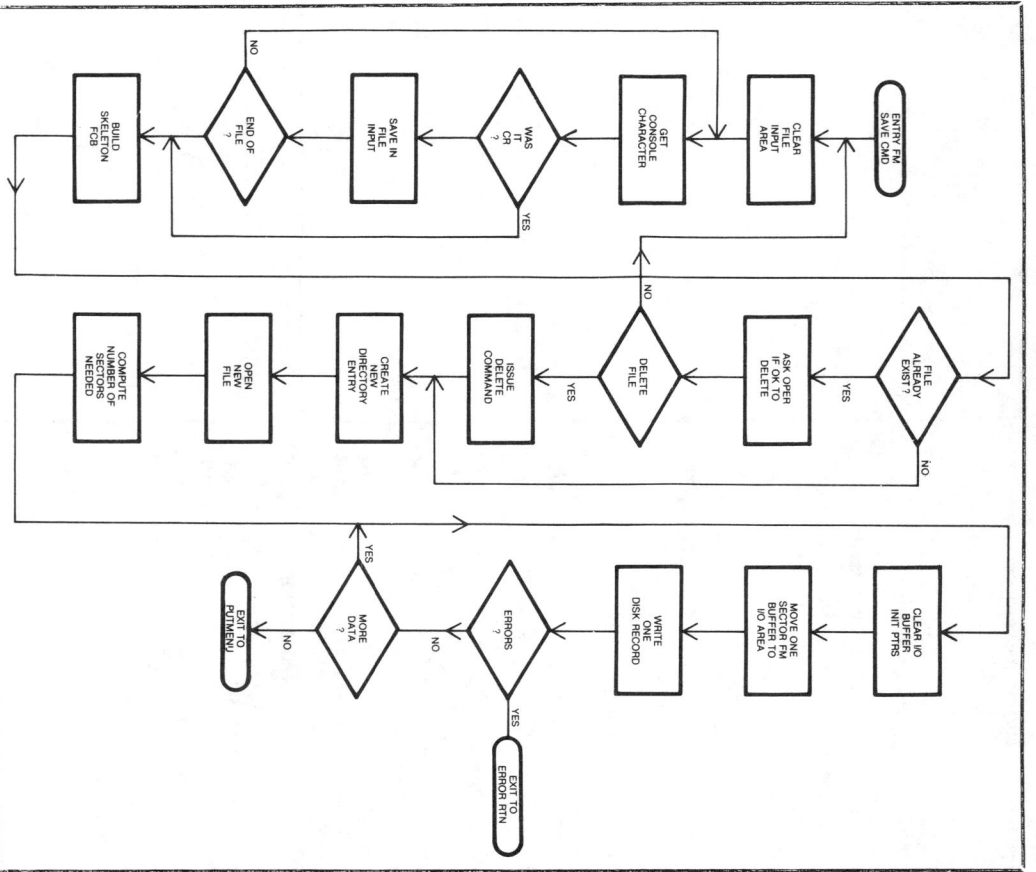

Fig. 11 — This routine takes the telemetry data in the computer buffer and saves it on disk.

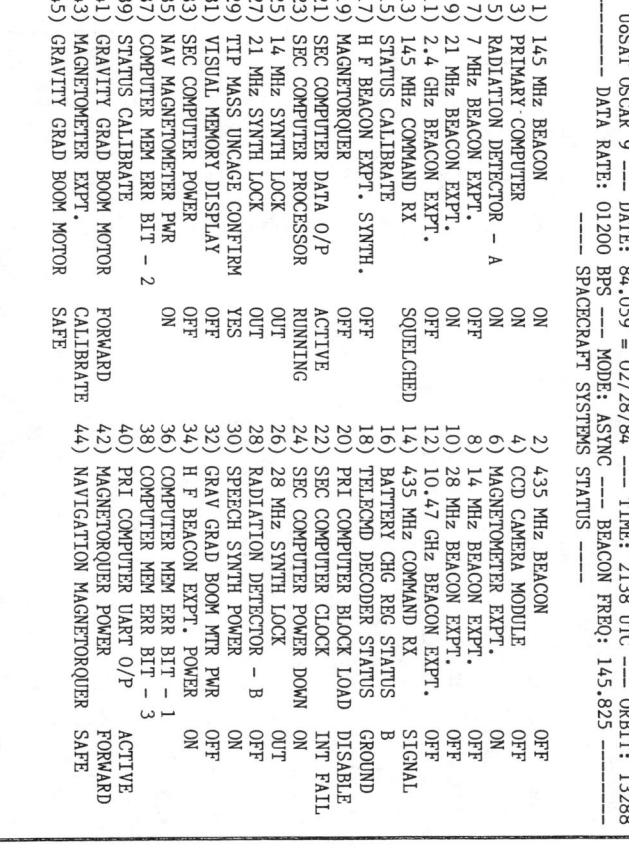

```
UoSAT OSCAR 9 ---- DATE: 84.059 = 02/28/84 ---- TIME: 2138 UTC ---- ORBIT: 13288
---- DATA RATE: 01200 BPS ---- MODE: ASYNC ---- BEACON FREQ: 145.825 ----
---- SPACECRAFT SYSTEMS STATUS ----

 1) 145 MHz BEACON                   ON         2) 435 MHz BEACON                   OFF
 3) PRIMARY COMPUTER                 ON         4) CCD CAMERA MODULE                OFF
 5) RADIATION DETECTOR - A           ON         6) MAGNETOMETER EXPT.               ON
 7) 7 MHz BEACON EXPT.               OFF        8) 14 MHz BEACON EXPT.              OFF
 9) 21 MHz BEACON EXPT.              ON        10) 28 MHz BEACON EXPT.              OFF
11) 2.4 GHz BEACON EXPT.             OFF       12) 10.47 GHz BEACON EXPT.           OFF
13) 145 MHz COMMAND RX               SQUELCHED 14) 435 MHz COMMAND RX               SIGNAL
15) STATUS CALIBRATE                            16) TELECMD DECODER CHG REG STATUS  B
17) H F BEACON EXPT. SYNTH.          OFF       18) PRI COMPUTER BLOCK LOAD          GROUND
19) MAGNETORQUER                     OFF       20) SEC COMPUTER CLOCK               DISABLE
21) SEC COMPUTER DATA O/P            ACTIVE    22) SEC COMPUTER POWER DOWN          INT FAIL
23) SEC COMPUTER PROCESSOR           RUNNING   24) SEC COMPUTER POWER               OFF
25) 14 MHz SYNTH LOCK                OUT       26) 28 MHz SYNTH LOCK                OUT
27) 21 MHz SYNTH LOCK                OUT       28) RADIATION DETECTOR - B           OFF
29) TIP MASS UNCAGE CONFIRM          YES       30) SPEECH SYNTH POWER               OFF
31) VISUAL MEMORY DISPLAY            OFF       32) GRAV GRAD BOOM MTR PWR           ON
33) SEC COMPUTER POWER               OFF       34) H F BEACON EXPT. POWER           ON
35) NAV MAGNETOMETER POWER           ON        36) COMPUTER MEM ERR BIT - 1         ON
37) COMPUTER MEM ERR BIT - 2                   38) COMPUTER MEM ERR BIT - 3
39) STATUS CALIBRATE                 FORWARD   40) PRI COMPUTER UART O/P            ACTIVE
41) GRAVITY GRAD BOOM MOTOR                    42) MAGNETORQUER POWER               FORWARD
43) MAGNETOMETER EXPT.               CALIBRATE 44) NAVIGATION MAGNETOMETER          SAFE
45) GRAVITY GRAD BOOM MOTOR          SAFE
```

Fig. 12 — An example of the decoded spacecraft systems status. The data shown here are decoded from the lines beginning with "AMSAT" as shown in the raw telemetry sample (Fig. 1) for UoSAT OSCAR-9.

the diskette, what can be done to improve its integrity? Several things, and these are accomplished during the editing phase. The edit phase simply reads the captured data and writes a new file containing only error-free records. Some items that can be checked during the editing phase are

1) The length of the telemetry lines that were saved. If any are of incorrect length, the whole line can be discarded.

2) Proper line data. Are the lines spaced properly? Are frame numbers ascending and between the proper limits? Do the values within the lines make sense? Do the checksum calculations yield the proper result?

It is impossible to detect every kind of error, but a good editing job will save you headaches later. You should also add some type of indication as to when the data were collected. (See Figs. 12 and 13.) I add a header to the output file. The header contains the satellite name, data, time, data rate, orbit number, beacon frequency and transmission mode.

The analysis phase consists of reading the edited telemetry file and substituting the values into the calibration equations. You can collect data over a long period of time and then produce graphic displays for that period.

Summary

This information should provide a starting point for those of you who would like to make a permanent record of the data being transmitted by UoSAT-OSCAR 9. Even though the system I described is dependent on the hardware in my computer system, I hope you will be able to apply the principles shown here to your own computer system.

```
UoSAT OSCAR 9  ---  DATE: 84.059 = 02/28/84  ---  TIME: 2138 UTC  ---  ORBIT: 13288
---  DATA RATE: 01200 BPS  ---  MODE: ASYNC  ---  BEACON FREQ: 145.82  ---
---  SPACECRAFT TELEMETRY  ----  1  ---

CHANNEL  PARAMETER                           RAW VALUE   ACTUAL       UNITS
00       SEC COMP CURRENT                    110         132.000      mA
01       SOLAR ARRAY CURRENT +X              020         222.400      mA
02       BATTERY HALF VOLTAGE                773           7.807      Volts
03       RADIATION DETECTOR A O/P            001          41.600      Count
04       RADIATION DETECTOR B O/P            001          41.600      Count
05       MAGNETOMETER HY-COARSE              528        1176.150      nT
06       MAGNETOMETER HY-COARSE              529        2837.620      nT
07       MAGNETOMETER HZ-COARSE              713       27425.480      nT
08       BATTERY PACK-A TEMP                 458           3.232      Degrees C
09       +X FACET TEMP                       472           0.404      Degrees C
10       VISUAL DISPLAY & CCD CURRENT        100          84.000      mA
11       SOLAR ARRAY CURRENT +Y              150         368.000      mA
12       2.4 GHz BEACON POWER O/P            000           0.000      mW
13       RADIATION DETECTORS EHT VOLTS       370         370.000      Volts
14       RADIATION DETECTORS CURRENT         307          40.180      nT
15       MAGNETOMETER HX-FINE                628        2111.850      nT
16       MAGNETOMETER HY-FINE                564         970.380      nT
17       MAGNETOMETER HZ-FINE                537         479.520      nT
18       BATTERY PACK-B TEMP                 393          16.362      Degrees C
19       -X FACET TEMP                       416          11.716      Degrees C
20       PRI COMP CURRENT                    160         162.000      mA
21       SOLAR ARRAY CURRENT -X              200         424.000      mA
22       BATTERY/BCR 14V BUS                 715          15.101      Volts
23       SUN SENSOR +Z AXIS                  112           0.566      Volts
24       10.4 GHz BEACON CURRENT             008          -7.760      mA
25       MAGNETOMETER TEMP                   419          11.110      Degrees C
26       MAGNETOMETER CURRENT                439          54.573      mA
27       TELECOMMAND RX CURRENT              283          31.773      mA
28       RADIATION EX TEMP +X1               468           1.212      Degrees C
29       +Y FACET TEMP                       589         -23.230      Degrees C
30       BATTERY CHARGE CURRENT              310         930.000      mA
31       SOLAR ARRAY CURRENT -Y              390         636.800      mA
32       POWER COND MODULE +10V              667          10.338      Volts
33       TELEMETRY SYS CURRENT               256           8.672      mA
34       2.4 GHz BEACON CURRENT              004          -3.002      mA
35       145 MHz BEACON POWER O/P            357         459.250      mW
36       145 MHz BEACON CURRENT              395          98.358      mA
37       145 MHz BEACON TEMP                 358          23.432      Degrees C
38       PRI COMP TEMP -X1                   483          -1.818      Degrees C
39       -Y FACET TEMP                       220          51.308      Degrees C
40       +14 V LINE CURRENT                  090         257.400      mA
41       +5 V LINE CURRENT                   100          64.000      mA
42       POWER COND MODULE +5V               739           5.518      Volts
43       SUN SENSOR -Z AXIS                  036           0.182      Volts
44       HF BEACONS CURRENT                  139          35.638      mA
45       435 MHz BEACON POWER O/P            000           0.000      mW
46       435 MHz BEACON CURRENT              003         -10.881      mA
47       435 MHz BEACON TEMP                 424          10.100      Degrees C
48       SEC COMP TEMP +Y1                   505          -6.262      Degrees C
49       +Z FACET TEMP                       458           3.232      Degrees C
50       +10V LINE CURRENT                   100         300.000      mA
51       -10V LINE CURRENT                   090          39.000      mA
52       POWER COND MODULE -10V              275           4.113      mA
53       NAV MAGNETOMETER Y-AXIS             115     -100631.062      nT
54       NAV MAGNETOMETER Z-AXIS             863      -99783.333      nT
55       NAV MAGNETOMETER X-AXIS             299       38423.625      nT
56       SPEECH SYNTH CURRENT                462          45.001      mA
57       CCD IMAGER TEMP                     469           1.010      Degrees C
58       TELEMETRY SYS TEMP -Y1              441           6.666      Degrees C
59       -Z FACET TEMP                       477          -0.606      Degrees C
```

Fig. 13 — A sample decoded telemetry frame. The channel numbers 00-59 correspond to the first two digits in the 5-digit telemetry groups. The values in the column RAW VALUE are the other three digits from each channel. These are substituted into the proper calibration equation by the analysis program, and the values shown in the right-hand column result.

A Profile of the UoSat-OSCAR 11 Satellite

By Jon Bloom, KE3Z
ARRL Senior Laboratory Engineer

The spacecraft named "UoSat" were built by the UoSat Spacecraft Engineering Research Unit of the Department of Electronic and Electrical Engineering of the University of Surrey, England. At this writing, there are two active satellites in the UoSat series: UoSat-OSCAR 9 (UoSat 1) and UoSat-OSCAR 11 (UoSat 2). Each of these satellites was built in order to support scientific experimentation and space education and to engender a low-cost-satellite design and construction capability at Surrey. UoSat 1 is described in the ARRL publication, *The Satellite Experimenter's Handbook*.

UoSat 2: A Second-Generation Experiment

UoSat 2 was launched aboard a Delta rocket on March 1, 1984, from Vandenberg Air Force base in California. The polar, sun-synchronous orbit is nearly circular at an altitude of approximately 680 km (423 miles). UoSat 2 circles the earth approximately once every 99 minutes.

In addition to the spacecraft "housekeeping," such as the telemetry and attitude control systems, UoSat 2 contains a number of *experiments*. The Digital Communication Experiment (DCE) provides a mechanism for sending and receiving digital data. An NSC-800-based microcomputer aboard the satellite controls this experiment and includes 128 kbytes of memory. The DCE is configured as a "flying mailbox," allowing messages to be stored and retrieved by authorized ground stations.

The Space Dust Experiment, built by students at the University of Kent, England, detects impacts of dust particles on the experiment sensors and measures the momentum of the particles.

The Charge-Coupled-Device (CCD) Camera Experiment is a redesigned version of an experiment flown on UoSat 1. It produces a picture of 384 by 256 pixels, each of which can take on any of 128 brightness levels. The camera points toward the earth.

The Digital Store and Readout (DSR) Experiment stores the information from the CCD and space dust experiments along with data from the on-board computer and

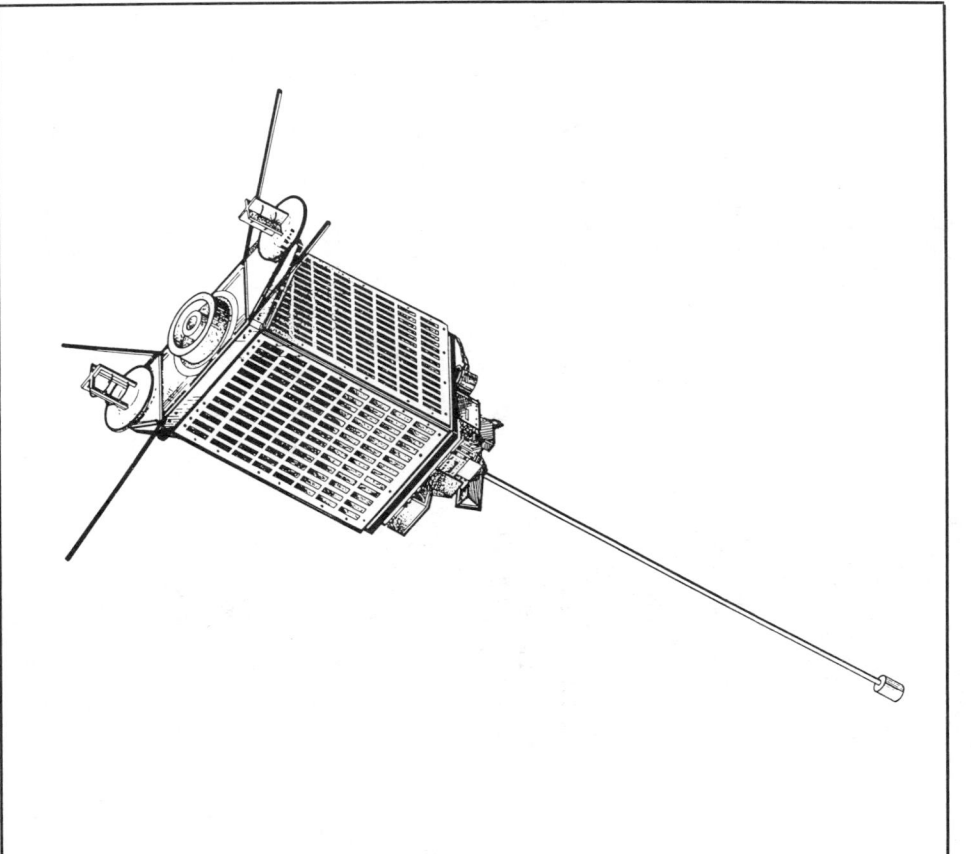

formats the data for transmission.

The UoSat 2 Ground Station

Like UoSat 1, UoSat 2 uses Amateur Radio primarily as a means to an end, rather than an end in itself. That is, the transmissions from UoSat-2 are performed within the amateur bands in keeping with the experimental nature of the satellite. The frequencies on which UoSat transmissions may be received are 145.825, 435.025 and 2401.5 MHz. The signals carry digital data transmitted using audio frequency-shift keying (AFSK) on a standard FM signal. (Note: the 435.025- and 2401.5-MHz signals can be phase-shift keyed as well.) When the satellite signals are strongest (on orbits when the satellite passes nearly overhead at the receiving station), the signals can be heard on a minimal receiver—a hand-held transceiver with a "rubber duck" antenna. For serious users, a circular polarized (left-hand circular polarization) directional antenna is recommended.

To make use of the signals from UoSat 2, two special-purpose devices are needed: a *demodulator* and a computer or terminal. The demodulator changes the AFSK tones from the receiver audio output into the signal needed by the computer or terminal. The UoSat 2 signals are sent using 1200- and 2400-Hz tones, with a 1200-Hz tone representing a binary 0 (space) and a 2400-Hz tone representing a 1 (mark). Each 7-bit character of the data is sent in a serial frame with one start bit, one parity bit and two stop bits, at 1200 bauds. This format is compatible with most terminals and computers. Suitable demodulators for UoSat 2 have been described in several publications (and earlier in this anthology).[1,2] (Note: UoSat 1 uses the same transmission format except the tones are reversed. Demodulators designed for UoSat 1 must have their outputs inverted for use with UoSat 2.)

UoSat 2 Data Formats

A typical UoSat 2 telemetry data *frame* is shown in Fig 1. The data is preceded by an ASCII RS ($1E_{16}$) character. The first line of the frame identifies the source as UoSat 2 and gives the date and time. The following lines give numeric values for each *channel* of the telemetry (typically, a channel is the readout value from a single sensor on the spacecraft). The first 60 channels (channels 0 through 59) are representations of analog measurements. These values may be converted to standard units by use of the conversion formulas given in Table 1. The final 10 channels (60 through 69) are in hexadecimal and are decoded bit by bit (see Table 2).

As an example of telemetry conversion, channel 00 in Fig 1 has a received telemetry value of 504. Substituting this value into the telemetry equation for channel 00 from Table 1 gives:

$$I = 1.9 (516 - 504) = 22.8 \text{ mA}$$

UoSat 2 can also send data in other formats. *Whole Orbit Data* (WOD) are telemetry readings stored in the spacecraft during the orbit, when the spacecraft is out of view of a ground control station. Upon command from the ground station, the WOD can be sent in a specialized telemetry transmission. Computer status messages and data from the CCD experiment can also be transmitted.

The UoSat 2 DCE software, developed by Harold Price, NK6K, Jeff Ward, K8KA, and Larry Kayser, WA3ZIA/VE3, provide a message store-and-forward system. The intent of the system is that a few authorized stations act as *gateways* between terrestrial packet-radio networks and the satellite. This makes UoSat 2 an effective means of interconnecting distant packet networks for message traffic.

Periodically, the downlink transmission switches to the DCE, at which time frames are transmitted that tell a receiving station what messages are in the satellite, by addressee. A ground station can then establish communication with the satellite and read any messages that it can gateway to the local packet network.

UoSat 2 can also send computer-generated voice messages on the downlink via the Digitalker experiment. As this is being written, the Digitalker is being used to periodically send voice messages giving the progress of the Skitrek team across the Arctic.

For a detailed description of the UoSats, the *UoSat Spacecraft Data Booklet* can be obtained from:

UoSat Spacecraft Engineering Research Unit
Department of Electronic & Electrical Engineering
University of Surrey
Guilford, Surrey, GU2 5XH
ENGLAND

for GBP3.50 or $7.00, plus airmail postage (no US checks accepted).

Notes

[1] J. Miller, *Wireless World*, May 1983.
[2] R. Diersing, "Microcomputer Processing of UoSat-OSCAR 9 Telemetry, Aug 1984 *QST*, p 23.

```
UOSAT-2         8510270104138
005041014B2F026721033490040523050309F0602510705200804780903370
102982113322120003130637141253154415161B1F17514618482719539370
204215211B392266002300012400062500072609702755722B5130295248
3051343104063228603357BA3400073526643631B9F37430338476E395048
4076634112064264306324416704500014600024794A48506F4947799
50569F51102752677153684C546531550000560035749965849445907E
608266A615BE7621F4E633305644402651705664B2B67700668000E69000F
```

Fig 1—A UoSat 2 Telemetry Data Frame. Each telemetry channel is sent as 5 characters followed by a checksum character. The first two characters are the channel number; the next three characters are the telemetry channel. For analog channels (00 through 59) these are decimal digits. For digital channels (60 through 69) the digits are hexadecimal. For example, the data for channels 00 and 61 appear above as: 005041 and 615BE7

Table 1
UoSat 2 Analog Telemetry Channel Conversion

To calculate the analog value, replace N with the received 3-digit telemetry value.

Channel	Parameter	Equation
00	Solar array current −Y	I = 1.9 (516−N) ma
01	Nav mag X axis	H = (0.1485N − 68) μT
02	Nav mag Z axis	H = (0.1523N − 69.3) μT
03	Nav mag Y axis	H = (0.1507N − 69) μT
04	Sun sensor 1	N uncalibrated
05	Sun sensor 2	N uncalibrated
06	Sun sensor 3	N uncalibrated
07	Sun sensor 4	N uncalibrated
08	Sun sensor 5	N uncalibrated
09	Sun sensor 6	N uncalibrated
10	Solar array current +Y	I = 1.9 (516−N) mA
11	Nav mag (Wing) temp	T = (330 − N)/3.45 C
12	Horizon sensor	N uncalibrated
13	435 MHz Beacon VCO control	V = N/200 V
14	DCE RAMUNIT current	I = (N − 70.4)/6.7 mA
15	DCE CPU current	I = (N − 187.1)/2.0 mA
16	DCE GMEM current	I = (N − 121.3)/2.1 mA
17	Facet temp +X	T = (480 − N)/5 C
18	Facet temp +Y	T = (480 − N)/5 C
19	Facet temp +Z	T = (480 − N)/5 C
20	Solar array current −X	I = 1.9 (516−N) mA
21	+10V line current	I = 0.97N mA
22	PCM voltage +10V	V = 0.015N V
23	PW logic current (+5V)	I = 0.14 mA (N <= 500)
24	PW Geiger current (+5V)	I = 0.21N mA
25	PW Elec sp. curr (+10V)	I = 0.096N mA
26	PW Elec sp. curr (−10V)	I = 0.093 mA
27	Facet temp −X	T = (480 − N)/5 C
28	Facet temp −Y	T = (480 − N)/5 C
29	Facet temp −Z	T = (480 − N)/5 C
30	Solar array current +X	I = 1.9 (516 − N) mA
31	−10V line current	V = 0.48N mA
32	PCM voltage −10V	V = −0.036N V
33	1802 comp curr (+10V)	I = 0.21N mA
34	Digitalker current (+5V)	I = 0.13 mA (N <= 500)
35	145 MHz beacon power O/P	P = (2.5N − 275) mW (N > 200)
36	145 MHz beacon current	I = 0.22N mA
37	145 MHz beacon temp	T = (480 − N)/5 C
38	Command decoder temp (+Y)	T = (480 − N)/5 C
39	Telemetry temp (+X)	T = (480 − N)/5 C
40	Solar array voltage (+30V)	V = (0.1N − 51.6) V
41	+5V line current	I = 0.97N mA
42	PCM voltage +5V	V = 0.0084N V
43	DSR current (+5V)	I = 0.21N mA (N <= 500)
44	Command RX current	I = 0.92N mA
45	435 MHz beacon power O/P	P = (2.5N − 200) mW (N > 175)
46	435 MHz beacon current	I = 0.44N mA
47	435 MHz beacon temp	T = (480 − N)/5 C
48	PW temp (−X)	T = (480 − N)/5 C
49	BCR temp (−Y)	T = (480 − N)/5 C
50	Battery charge/dischg curr	I = 8.8 (N − 513) mA
51	+14V line current	I = 5N mA
52	Battery voltage (+14V)	V = 0.021N V
53	Battery cell volts (MUX)	I = N uncalibrated
54	Telemetry current (+10V)	I = 0.02N mA
55	2.4 GHz beacon power O/P	P = ((N+50)2)/480 mW
56	2.4 GHz beacon current	I = 0.45N mA
57	Battery temp	T = (480 − N)/5 C
58	2.4 GHz beacon temp	T = (480 − N)/5 C
59	CCD imager temp	T = (480 − N)/5 C

Table 2
UoSat 2 Digital Telemetry Channel Conversion

Channel	Bit	Parameter	State
60 (MSB)	1	145 MHz General beacon power	Off/On
60	2	435 MHz Engineering Beacon power	Off/On
60	3	2401 MHz Engineering Beacon power	Off/On
60	4	Telemetry channel mode select	Run/Dwell
60	5	Telemetry channel dwell address load	Off/On
60	6	Telemetry channel dwell address source	Gnd/Computer
60	7	Primary Spacecraft Computer power	Off/On
60	8	Primary Spacecraft Computer error count	Bit 1
60	9	Primary Spacecraft Computer error count	Bit 2
60	10	Primary Spacecraft Computer bootstrap	UART/PROM
60	11	Primary Spacecraft Computer error count	Bit 3
60 (LSB)	12	Primary Spacecraft Computer bootstrap	A/B
61 (MSB)	13	Gravity gradient boom deployment pyros	Safe/Arm
61	14	Gravity gradient boom deployment pyros	Fire/Hold
61	15	Gravity gradient boom deployment	Safe/Arm
61	16	Gravity gradient boom deployment	Deploy/Hold
61	17	Gravity gradient boom deployment	Extend/Retract
61	18	Attitude Control Magnetorquers –X	Safe/Arm
61	19	Attitude Control Magnetorquers –X	On/Off
61	20	Attitude Control Magnetorquers –Y	On/Off
61	21	Attitude Control Magnetorquers –Z	On/Off
61	22	Attitude Control Magnetorquers	Reverse/Forward
61	23	435 MHz PSK mode	NRZI/NRZIC
61 (LSB)	24	2401 MHz PSK mode	NRZI/NRZIC
62 (MSB)	25	Attitude Control Magnetorquers	High/Low power
62	26	Digitalker expt. power	Off/On
62	27	CCD Camera expt. power	Off/On
62	28	CCD Camera expt. integration period	Bit 0
62	29	CCD Camera expt. integration period	Bit 1
62	30	CCD Camera expt. video amp gain	Bit 0
62	31	CCD Camera expt. video amp gain	Bit 1
62	32	DSR power	Off/On
62	33	DSR mode	Read/Write
62	34	DSR mode	Run/Reset
62	35	Radiation Detectors Geiger-A EHT power	Off/On
62 (LSB)	36	Radiation Detectors Geiger-B EHT power	Off/On
63 (MSB)	37	Radiation Detectors Geiger-C EHT power	Off/On
63	38	Electron Spectrometer sensor EHT power	Off/On
63	39	DCE expt. power	Off/On
63	40	DCE expt.	Reset/Run
63	41	DCE expt. PROM select	A/B
63	42	DCE expt. CPU clock rate select	0.9/1.8 MHz
63	43	Navitagion Magnetometer power	Off/On
63	44	Space Dust experiment power	Off/On
63	45	Status calibrate	
63	46	BCR status	A/B
63	47	435 MHz beacon modulation select	AFSK/PSK
63 (LSB)	48	2401 MHz beacon modulation select	AFSK/PSK
64 (MSB)	49	Engineering data	Bit 1
64	50	Engineering data	Bit 2
64	51	Engineering data	Bit 3
64	52	Engineering data	Bit 4
64	53	Engineering data	Bit 5
64	54	Command Watchdog	Disable/Enable
64	55	Command Watchdog reset	Run/Reset
64	56	145 MHz beacon data select	A
64	57	145 MHz beacon data select	B
64	58	145 MHz beacon data select	C
64	59	145 MHz beacon data select	D
64 (LSB)	60	145 MHz beacon data select	E
65 (MSB)	61	145 MHz beacon data rate	F
65	62	145 MHz beacon data rate	A
65	63	145 MHz beacon data rate	B
65	64	435 MHz beacon data rate	A
65	65	435 MHz beacon data rate	B
65	66	435 MHz beacon data rate	C
65	67	Particle / Wavecounter control	Count/Reset
65	68	VHF/UHF beacon lockout protection	Disable/Enable
65	69	Engineering data	Bit 6
65	70	Engineering data	Bit 7
65	71	Engineering data	Bit 8
65 (LSB)	72	Engineering data	Bit 9
66 (MSB)	73	P/W channel plate control	Bit 2
66	74	P/W channel plate control	Bit 1
66	75	P/W channel plate control	Bit 0
66	76	Space Dust (MSB)	
66	77	Space Dust	
66	78	Space Dust	
66	79	Space Dust	
66	80	Space Dust	
66	81	Space Dust	
66	82	Space Dust	
66	83	Space Dust	
66 (LSB)	84	Space Dust (LSB)	
67 (MSB)	85	DSR write cycle complete	No/Yes
67 (MSB)	86	1802 CWO output	
67	87	1802 Telemetry port (MSB)	
67	88	1802 Telemetry port	
67	89	1802 Telemetry port	
67	90	1802 Telemetry port	
67	91	1802 Telemetry port	
67	92	1802 Telemetry port	
67	93	1802 Telemetry port	
67	94	1802 Telemetry port	
67	95	1802 Telemetry port	
67 (LSB)	96	1802 Telemetry port (LSB)	

Amateur Satellite Communications

Conducted By
Vern "Rip" Riportella, WA2LQQ
P.O. Box 177, Warwick, NY 10990

Antennas for Working OSCAR

It seems likely man has always looked skyward in search of inspiration. Today, when amateurs seek to expand their vistas, they naturally look skyward toward OSCAR. What antennas do they use?

A typical OSCAR station is equipped for three or perhaps four Modes: A, B, J and maybe L.[1] This implies the station has antennas for 10 m, 2 m and 70 cm. A fully equipped station may also include a 24-cm antenna for Mode L. What kind of antennas are typical on each band?

For 10-meter Mode A reception, there are several approaches. With a 10-meter beam and preamp, you'll hear the Russian amateur satellites well. However, when the RS satellites are overhead, your beam's low elevation pattern is a disadvantage. Supplement your 10-meter beam with another type of 10-meter antenna. Several selectable sloping dipoles help. A turnstile is easily built and surprisingly effective for overhead passes. A horizontal dipole stapled to a bedroom wall will even suffice, but a vertical is generally a poor choice.

Two-meter OSCAR antennas are used for Modes A, B, J and the future JL mode. Antennas for VHF and above can be interesting and occasionally baffling.

Recent OSCARs, such as AO-10, use circular polarization (CP). CP can be thought of as a mixture of vertical and horizontal polarization. AO-10 is spinning and what might be vertically polarized one moment could be horizontal the next. Moreover, radio waves passing through the geomagnetic field have their polarization rotated in passing. This is called Faraday rotation. The polarization of the downlink at your QTH is not only unpredictable, but it's rapidly varying! A CP antenna solves this problem by (ideally) being equally effective in both horizontal and vertical planes.

CP comes in two "flavors" (engineers call them senses): Right Hand (RHCP) and Left Hand (LHCP). A handy visualization tool for CP is to imagine you could see the plane waves launched by your CP antenna. Imagine a ribbon-like plane wave twisting off regularly into space. The rate of twist of the ribbon would be exactly one turn per wavelength of travel. The direction of the twist is the sense. If a mouse were to run down the RHCP ribbon (away from you), the mouse would appear to be rotating clockwise as he scampered off into the distance. Conversely, a mouse running down a LHCP ribbon would appear to be rotating counterclockwise.

Many antennas can generate CP, but few of them do it really well. Two CP figures of merit are gain and circularity (axial ratio). Circularity is the relative sensitivity of the antenna to RF energy in various planes. Gain for CP antennas is analogous to linearly polarized Yagis.

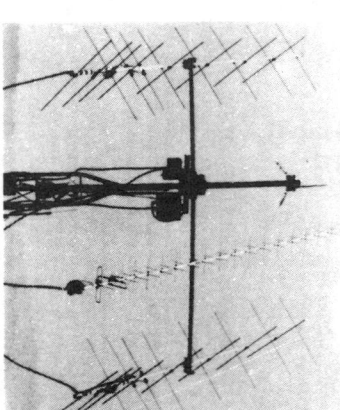

A typical Mode B OSCAR antenna system—a pair of 2-meter crossed Yagis and a single 70-cm crossed Yagi. See text for details.

The easiest way to get CP on 2 m is with a helical antenna. The 2-meter helix tends to be bulky, but it can be fun to build and is very forgiving of lax dimensional tolerances. Several types of helices are described in the *Satellite Experimenter's Handbook* (SEH).[2] Helices tend to be a little lower in gain than you might want for 2 meters.[3]

The most popular solution for the Mode B downlink is crossed Yagis. The Yagis can be on different booms or they may share a common boom. They are oriented at right angles to each other and fed so that the resultant wave acts like that imaginary ribbon. This can be done by equally splitting power to the two Yagis and then delaying the power to one Yagi by 90 degrees. The result will be either RHCP or LHCP, depending on which antenna is leading and which is lagging. An alternative is to have both Yagis on a common boom but to offset them physically on the common boom by ¼ wave. Then they can be fed in phase to generate either RHCP or LHCP. To reverse the sense, a ½-wave delay line is inserted in line with one or the other. AO-10 uses RHCP for all its antennas except the 2-meter omni, which is rarely used. Experience has shown it very advantageous to be able to switch your station antennas between the two senses (RHCP to LHCP).

OSCAR antennas for 70 cm are currently used for Modes B and L. In the future, they will be used for Modes JL and S. Helices and crossed Yagis are very popular here, too. Gain of 14 or 15 dBic can be obtained with a single crossed Yagi of modern design. You'll see the quagi antenna also used on 70 cm. Some work has been done on a CP version of a quagi which, in its original version, was linearly polarized.

For Mode L at 24 cm, Yagis, loop Yagis, quagis, helices, dishes and some other rare varieties enter the picture. F9FT (Tonna) has made an excellent Yagi array for Mode L, and many are using it. It produces linear polarization. Consequently, users may observe some spin modulation.[4] Currently experiencing a surge in popularity on Mode L are loop Yagis.[5] Some designers claim close to 20-dBi gain for the longer versions. Arrays are in use producing 24 dBic or more. KØRZ has had excellent results with an array of four long-loop Yagis for Mode L. Spectrum International sells a shorter version that seems to work quite well. These antennas are very lightweight, have low wind drag and are relatively easy to construct. Dimensional tolerances need to be precisely held however. The loop Yagis are also linearly polarized and show some spin modulation effects. Also used at 24 cm are small- to medium-sized parabolic dish antennas of the 4- to 8-foot class. (We'll cover these in a subsequent column.)

Let's take a quick look at a fairly typical Mode B OSCAR antenna installation. The photo shows a pair of 2-meter crossed Yagis and a single 70-cm crossed Yagi. The 2-meter Yagis are spaced at about 9 feet. The 70-cm Yagi is more than a ½ wavelength (at 2 meters) from the 2-meter antenna. The elements of the 70-cm crossed Yagis are at 45 degrees to the 2-meter elements to improve isolation. The boom is off-center to increase the distance of the 70-cm antenna from the mast. Top guys are used to lower mechanical stresses on the elevation rotor caused by the off-center arrangement. The top guys are of nylon 150-pound-test line with turnbuckles. The feeds for the antennas all approach from the rear to avoid pattern distortion caused by running the coax and its associated field through the elements. The boom is, of course, nonconductive fiberglass. This is a must! The preamps for each band are mounted in the two black boxes just below the elevation rotor.

When Galileo first turned his telescope on the skies, a whole new world opened to him. What will *you* find when you turn your antennas skyward?[6]

Next time we'll discuss ways of keeping your OSCAR antennas accurately pointed: tracking.

Notes

[1]Transponder modes were delineated in Table 1 in the September 1985 column.
[2]Available from ARRL and AMSAT.
[3]For 2-meter OSCAR work, you want at least 12-dBic gain.
[4]Spin modulation is the received signal amplitude modulation that results from the spin of the satellite and the consequent satellite antenna pattern rotation. The effect is akin to a low-frequency (approximately 100 Hz) amplitude modulation impressed on signals.
[5]J. Reisert, W1JR, "VHF/UHF World," *Ham Radio*, Sept. 1985, pp. 56-62.
[6]Information about getting started on OSCAR is available for an s.a.s.e. to the author at the address above.

Amateur Satellite Communications

Conducted By
Vern "Rip" Riportella, WA2LQQ
P.O. Box 177, Warwick, NY 10990

Basic Satellite-Tracking Themes

Establishing and maintaining knowledge of the whereabouts of satellites is called tracking.[1] There's a keen interest in tracking OSCAR. And I think I know why.

It's the mystery that shrouds the movements of these artificial planets that makes their movements so fascinating, their reappearance almost mystical. It makes those who accurately predict their appearance rather shaman-like figures in the community.

Most satellite shamans have replaced their icons with IBM®, their crow's feet with Commodore®, their mushrooms with MacIntosh® PCs. The great caldron is first loaded with Kepler stew and feverishly stirred. Soon it yields up a magnificent information porridge richer by far than any gypsy's tea leaves: a forecast of heavenly events to come ... the rise of a great friendly star, OSCAR!

There are two fundamental reasons one needs to keep track of OSCARs. First of all they move—some fast, others not so fast. Since you must use a beam of some sort, you need to know where to point to catch the little bugger as it goes by. Second, most folks would find it rather onerous to sit by a radio and wait several hours for a satellite to show up. That would be an awful waste of time, and could be hazardous to your health besides. The soothing effect of the white noise emanating from the receiver has been known to induce a condition known to medicine as radiosomnia: nodding off at just the wrong moment.[2]

To avoid entirely this potentially debilitating condition, one needs to know when the satellite of interest will appear. Then, one can leisurely go about one's (presumably) less somniferous business to return to the shack at just the right moment to catch the bird on the rise.

Thus, there are two fundamental functions of your OSCAR tracking effort: position determination and scheduling. There are other functions one might envision (and those will be discussed subsequently), but these two functions are clearly most basic.

To track OSCAR requires four things:

1) You need information about the OSCAR you want to track: its precise location and rates of movement at a precisely defined instant.
2) You must know your location.
3) You must know the time of day reasonably accurately.
4) Most important, you need a contrivance to make sense of the first three items.

The contrivance, or tool, can be an OSCARLOCATOR, Satellipse or similar manual tracking device.[3] Or it can be a number engine: a modern-day icon, crow's foot or mushroom. The manual locators are sufficiently accurate for most applications, and can cost under $20. Obviously, the crow's feet are somewhat more dear (ask any crow).

Let's look first at the manual locators. In the usual arrangement, you get an equidistant azimuthal projection map centered on the North Pole. A series of clear-plastic overlays is supplied. The ground-track overlay relates the path of the satellite to the map of the earth. The ground-track overlay pivots around a rivet positioned at the North Pole. In this way, it denotes various paths the satellite can take. Another overlay, called a spiderweb because of its resemblance to same, provides azimuth and elevation pointing information to the satellite from your QTH.[4] Time hacks on the ground-track overlay signify where the satellite will be at selected times after a reference is crossed. The satellite's motion (rates and angles) are factored into your plotter and determine the shape of the curves on the overlays. Different satellites usually have different orbits. That means you need a separate ground-track overlay for each satellite.[5]

To make the manual plotters work you need an accurate reference point, a precise reckoning of where the satellite was at a selected instant. With low earth-orbiting (LEO) satellites, which have very nearly circular orbits, the reference used is the time and longitude of the satellite's northbound equator crossing. More precisely, the reference orbit (the one most often used for reference purposes) is the first northbound equator crossing of the UTC day.

The reference orbit consists of the time and longitude of the equator crossing. Orbit number is sometimes also included. To track subsequent orbits you increment the longitude of the reference orbit by a fixed amount (given in the bulletins). Similarly, the time increment is given. So, given a reference orbit together with the longitude and time increments (orbital period), you have sufficient information for accurate tracking for several weeks, at least.

With elliptical orbiting satellites such as AMSAT-OSCAR 10 (AO-10), things are a bit more complex. Since both the velocity and altitude of AO-10 change constantly, the shape of the ground-track overlays seems at first a bit bizarre. The earth is rotating at a fixed rate, but the motion of the satellite is variable. The interaction of the two motions yields a ground track that doubles back on itself. Moreover, the shape of the ground-track overlay must be updated periodically to account for changes in the orbit geometry. Finally, when used for AO-10, the manual trackers most often use the time and location (both latitude and longitude) of apogee as a reference rather than the equator crossing time.[6]

Locators used for both circular and elliptical orbits have several things in common: They depend on well-defined graphs of the satellites' movements, and they require a reference time and position input to get started tracking. These reference inputs can be obtained from several sources. W1AW transmits reference orbits as part of its regular bulletin regime. AMSAT nets carry extensive orbital information, including the reference orbits and reference apogees.[7] Amateur Satellite Report (ASR) also contains the required information.[8]

So far, we've looked at two of the four items mentioned above that you need. We've looked at the position and motion information you need, and we've mentioned some form of tool (contrivance) to garner some meaning from the information. Now for the easiest two requirements: your location and the precise time.

You can determine your location from an atlas or a good road map. You should be able to determine your location easily to within a tenth of a degree of both longitude and latitude. If you have difficulty, try calling your municipal airport for help. As for time of day, the National Bureau of Standards station WVV can be tuned in on 2.5, 5, 10, 15 and 20 MHz whenever conditions allow. Time hacks are given every minute. You should have an accurate clock, preferably digital, closely synchronized with WWV.

Manual locators have much to recommend them. They're inexpensive and completely adequate for most tracking applications. In fact, they are fun to use, if a bit confusing at first to the newcomer. Designers of AO-10 trackers (K2ZRO, W2GFF and K2UBC) spent years working out the details to produce some excellent, affordable instruments.

Expert DXers know where and when to point their beams, based on years of experience. Tracking satellites is similar in some ways, yet starkly different in others. It does take experience to become a proficient satellite tracker and to really understand what is going on. In this way satellite trackers and DXers are similar. On the other hand, DXers have to shrug their shoulders at the vagaries of the F2 mode when overwhelmed by too many variables; simple unpredictability. Thankfully, predicting OSCAR access is much, much more precise. There is enormous satisfaction in dumping a bunch of numbers into a computer, pushing a few buttons and being presented with a piece of paper that says an object traveling at 18,000 miles per hour is going to pop over your horizon in precisely 38 minutes and 22 seconds. And then it does! Shamans, man your caldrons!

Next time, we'll conclude the tracking theme with a look at the icons, crow's feet and mushrooms (computer number crunching) tracking methods.[9]

Notes

[1]Tracking in a general sense can mean all those measures one employs to locate the object of interest. In a narrower sense, tracking, as with radar, can involve an exchange of information, either cooperatively or passively, between the tracker and the target. When we speak of "tracking" OSCARs, we generally mean the broader sense of the word. Tracking OSCAR most often reduces to predicting its position in space and time as discussed (in the text) rather than exchanging location information with it.

[2]Symptoms include a red mark on the forehead caused by frequent, percussive contact with the operating position.

[3]See the June 1985 column.

[4]See! There really are mystical allusions here.

[5]For satellites of the same family, however, you can use a single overlay. For example, for RS-5, RS-7 and RS-8 (all the currently operating Russian satellites), a single ground-track overlay is sufficient. Separate overlays are required for UoSAT-OSCAR 9 and UoSAT-OSCAR 11 because of their disparate altitudes.

[6]Apogee is the highest point in a satellite's orbit; the point at which it's farthest from earth.

[7]Net times and frequencies were given in the June 1985 column.

[8]See the June 1985 column.

[9]A free catalog of AMSAT Software Exchange tracking programs is available for an s.a.s.e. to the author at the address above.

Amateur Satellite Communications

Conducted By
Vern "Rip" Riportella, WA2LQQ
P.O. Box 177, Warwick, NY 10990

Basic Satellite-Tracking Themes—Part 2

Last month, we talked of manual satellite tracking (locator) systems. This month, we conclude the basic tracking discussion with a review of computer tracking methods.

Computer programs designed to track satellites usually have a common set of "guzintas" (inputs) and a variety of "comsoutas" (outputs). The inputs are almost always in the form of the NASA orbital prediction bulletin. The prediction bulletins provide precise data on the position of the satellite at a reference instant (epoch), and are velocities and accelerations for the satellite. Based on the values supplied, the well-designed tracking program can output a wealth of information.

The output can be presented on your TV or CRT monitor, it can be printed or it might even be within view of a given satellite concurrent with you.

There have been three generations of programs. The first generation simply gave you a tabular display (or printout), i. e., time, azimuth and elevation and SSP of the chosen satellite for a selected period. For example, the computer could figure the pointing angles for OSCAR 9 for the next four hours.

The second-generation programs added graphics. Those developed by W0SL, GM4IHJ, W6WNK and others presented you with a Mercator projection map of the earth. Depending on the program, you would see the relative position of one or more satellites projected on the map. Provided is a table that tells which way to point to "look" at the satellite.

The first of the third-generation programs does all of this and much more. A system developed by W5SXD and WB5CCJ is a remarkable combination of slick graphics and enormous flexibility. It provides tools for the satellite operations planners as well as users. And it runs with the 8087 math coprocessor, so it is very fast.

What to Expect

What can be done besides merely giving you pointing angles to the bird?

1) Tracking from your location: Provides azimuth and elevation from your QTH for a selected satellite if the satellite is above the horizon. Advanced programs allow you to track several satellites simultaneously. Some will allow you to adjust the time step size to better resolve motion. Others let you set the horizon value so you are alerted to satellites just below the horizon where they can occasionally be worked. Most programs will tell you how far along in its orbital path the bird has progressed on the current or anomaly (MA). This is displayed in terms of phase or mean anomaly (MA). This is important because some satellites, AO-10 in particular, operate in different modes according to where they are along their orbital path.

2) Scheduling: This allows you to select a broad time window and ask the computer to display each rise and set of the selected satellite within that window. More advanced programs allow scheduling of more than one satellite.

3) Mutual visibility: Time and pointing information for any two stations you select that will be simultaneously within view of the chosen satellite.

4) Satellite off-pointing angle: This new feature, this facility tells you how far off from your QTH the satellite's beam antenna is pointing. This is significant with AO-10 since it uses high-gain beams and occasionally the satellite's orientation is adjusted for better sun-angle. The algorithm for doing the off-pointing calculations was developed by Bob McGwier, N4HY, and is in several versions of his QUIKTRAK program.

5) Graphics: The addition of graphics distinguishes the second-generation tracking system from the first. Advanced graphics have handsome maps with detailed land mass outlines. The simplest programs will track but one satellite on the map. The best graphics capabilities I've seen are in the W5SXD/WB5CCJ GRAFTRAK II package. Here you can watch the satellite's SSP traverse the map (as with the others) or you can plot the "footprint" (coverage zones). You can even zoom in to get more detail. Or you can do a special spherical projection to see what the earth looks like from the satellite at that instant. This, combined with its number crunching speed, distinguish GRAFTRAK II as the first (and for the moment only one) of the third-generation programs.

6) Autotrak: K0RZ homebrewed one of the first systems whereby the computer controls the antenna position directly without human intervention. Although fewer than 1 in perhaps 500 satellite users today have autotrak systems running now, they will become very popular in the future.

7) Special features: GRAFTRAK II has a fast mode that allows you to watch the ground track and coverage zone move across the map at select-table high rates. This helps to plan access times to various QTHs you might want to chat with.

While tracking a satellite is a fascinating and instructive exercise, there are those who would just as soon treat the media as transparent and get on with the fun of QSOing. For them, computer-aided tracking and eventually auto-tracking is definitely the way to go. The ease of use and tremendous information content afforded, especially from those programs such as GRAFTRAK II with advanced graphics capability, make satellite use all the more enjoyable.

Next time we'll discuss the ultimate in easy-to-track satellites: AMSAT's plans for a fully geosynchronous satellite system, which just hangs in the sky with no apparent movement at all; just set your antennas and weld them in place. Next month: Phase 4 Satellites. The future is here![2]

Notes

[1]"The satellite's position is usually given in terms of the latitude and longitude of its sub-satellite point (SSP). The SSP is the point on the surface of the earth lying on a line connecting the center of the satellite and the center of the earth. In other words, the SSP is the point right "beneath" the satellite. The position information given may also include the elevation above the earth and the range to your QTH.

[2]A free catalog of AMSAT Software Exchange tracking programs is available from the author at the address above.

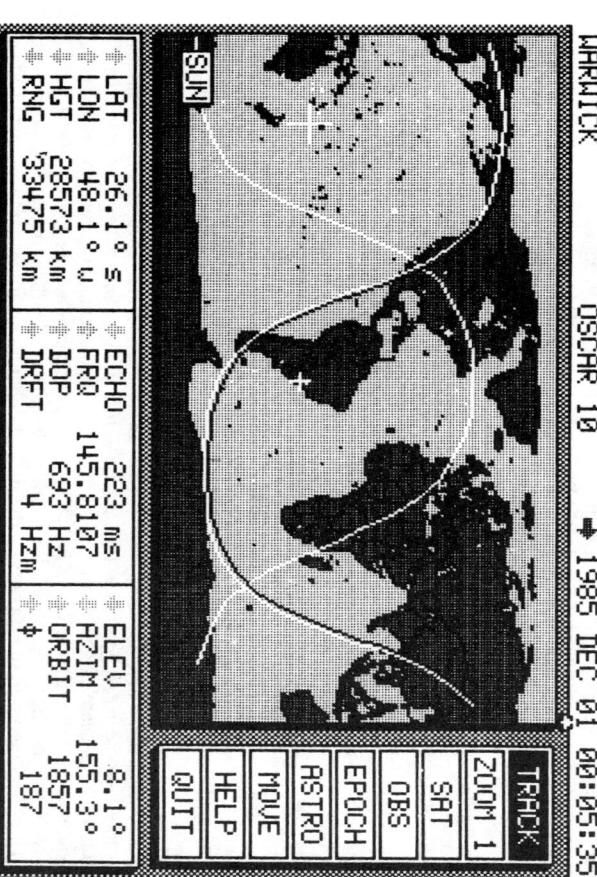

Examples of the newest and best graphics tracking abilities of the GRAFTRAK II software of Silicon Solutions (l-r); a footprint of AO-10 over South America and western Africa while the sun is centered north of New Zealand (left).

Satellite Anthology 58

Amateur Satellite Communications

Conducted By
Vern "Rip" Riportella, WA2LQQ
PO Box 177, Warwick, NY 10990

Fun, Games and (Hopefully) Technical Challenges on OSCARs

Viewed as "sport," Amateur Radio inspires some to reach new heights. But taking this "radiosport" competition to the OSCAR satellites raises both new challenges *and* risks. OSCAR transponders are especially vulnerable to the abuses intense competition seems to foster. Consequently, AMSAT has carefully limited on-satellite competition. Power "hogging" of limited transponder power by ardent contesters is the basic problem; it's the essence of what precludes "contests" on OSCARs.

Competition also means awards. Awards programs differ from contests in meaningful ways.

First, contests tend to be short, high-intensity events. Awards (eg, DXCC) tend to be of lower intensity and are usually long-term pursuits.

Second, contests pit operators against each other in a dueling environment. Awards present arbitrary milestones (eg, 100 countries). The absence of head-to-head combat seems to reduce abuse levels.

There is a third class of competition, which I call techno-sport. It lends itself nicely to satellites and embraces the essence of the Amateur Space Program. Techno-sport emphasizes technical acumen over physical endurance, mind-food over monastic chanting and ken over kilowatts! The now-defunct Frequency Measuring Test (FMT) and the Fox 'n' Hound (F'n'H) hidden-transmitter events are familiar examples.

The FMT challenged the participant to determine precisely the frequency of a CW signal. The more accurate the judgment, the higher the award. Today's F'n'H events pit clever foxes against portable, sophisticated equipment and are more fun than ever.

The desire to instill sound engineering principles and advanced techniques through the incentive of competition, combined with its fundamentally benign character, is the motivation for AMSAT's developing a series of techno-sport competitions. The first techno-sport competition held on AMSAT-OSCAR 10 (AO-10) in 1985 was a receive-sensitivity test. The participant merely copied a moderately weak CW signal transmitted through AO-10. If the characters were copied correctly, the participant was awarded a certificate. If a still weaker signal was received, the competitor would receive an endorsement to the award. Each time a weaker signal was copied, a more prestigious endorsement would be granted. This form of competition is simple and technically stimulating, too. It promotes improved "ears," through better station engineering, essential in OSCAR use.

AMSAT's receive-sensitivity test is called the K2ZRO Memorial Station Engineering Award. In the ZRO Test, the participant monitors AO-10's downlink. A calibrated CW signal sends a five-digit number group in 10-wpm Morse code. The participant notes it. Then the test director reduces the signal power level by half (−3 dB). A different code group is sent. The participant notes it. Power is cut in half again. This process is repeated until nine different power levels are sent: Z_0 through Z8. The lowest level is 24 dB below the first signal (see Table 1).

Conducting the test is also a challenge. Fig 1 shows the equipment used in the transmitter chain. The power meter is used only to gauge general power levels. Its accuracy (±5%) is insufficient for the test. In the future, a directional coupler and a precision Boonton microwattmeter with digital readout will be used. The isolator stabilizes the impedance the attenuator output sees. Attenuator accuracy hinges on the maintenance of a 50-ohm line, and the accuracy of the entire test depends on the accuracy of the attenuator.

Choosing the test time is as important to the accuracy of the ZRO Test as the instrumentation. The critical factors are satellite location, perspectives and attitude (orientation). Test fairness must be assured in three ways:

1) Each participant must have the same downlink energy to work with (uniformity).
2) No test session should be easier than another (precision repeatability).
3) Increments must be close to −3 dB (accuracy).

To assure uniform illumination levels, the test session is scheduled at (or near to) apogee, the highest point on the satellite's orbit. During this time, the difference in path length (and path loss) between participants is minimal. Also, when the satellite is at apogee, the differences attributable to downlink antenna-beam variations are minimized.

To assure repeatability, the benchmark for each session is the observed strength of the satel-

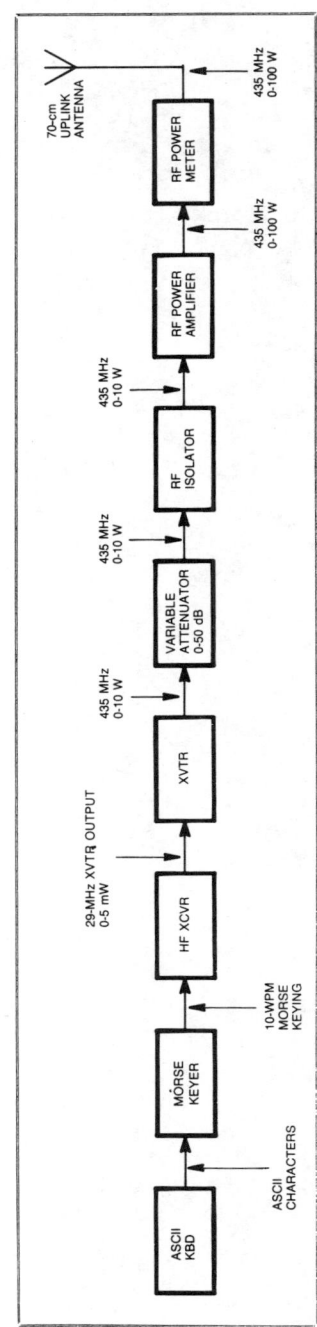

Fig 1—ZRO Test transmit chain.

Table 1
ZRO Test Typical Operation Levels

Z Level	Relative Level (dB)	Measured RF to Feed Line	EIRP*
0	0	20 W (13 dBW)	316 W (25 dBW)
1	−3	10 W (10 dBW)	158 W (22 dBW)
2	−6	5 W (7 dBW)	79 W (19 dBW)
3	−9	2.5 W (4 dBW)	40 W (16 dBW)
4	−12	1.25 W (1 dBW)	20 W (13 dBW)
5	−15	625 mW (−2 dBW)	10 W (10 dBW)
6	−18	312 mW (−5 dBW)	5 W (7 dBW)
7	−21	156 mW (−8 dBW)	2.5 W (4 dBW)
8	−24	78 mW (−11 dBW)	1.25 W (1 dBW)

*Antenna Effective Isotropic Radiated Power, assuming 1.0-dB feed-line loss and 13.0-dB gain.

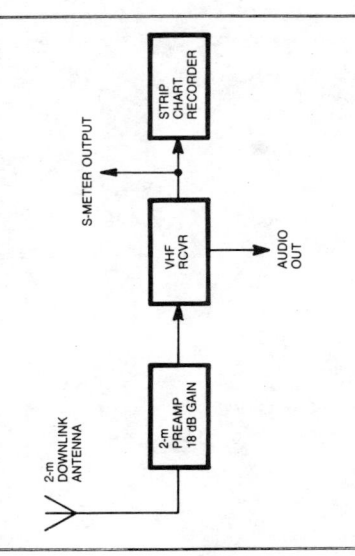

Fig 2—ZRO Test receive chain.

lite's beacon transmitter rather than, say, a carrier sent through the transponder. To assure accuracy of the −3 dB increments, a precision attenuator is used.

To calibrate a test shot, the test director carefully adjusts the uplink power such that the downlink signal exactly equals the observed beacon amplitude. This, then, becomes the reference Z_0 level for the test session (316 watts effective isotropic radiated power [EIRP] uplink in the session typified in Table 1).

Setting the Z_0 level is often difficult; it is the reason for using the strip chart recorder in the receiver chain of Fig. 2. With spin modulation and variable transponder loading to complicate matters, careful observation is required before the baseline Z_0 level can be established.

Choice of the transponder test environment is essential to the success of the test session. The transponder loading must be moderate and stable. Too few users means statistically unacceptable variations in transponder loading such that a single, strong signal could modulate the test levels. Too many users reduces the effect of individual users, but may mean QRM and demand more uplink power than the power amplifier can provide in its linear operating range.

Once the benchmark Z_0 is set, it is simple to reduce power levels in 3-dB increments using an attenuator. One of the significant results to date is that it is *much* easier to attain Z8 on Mode L (70-cm downlink) compared to Mode B (2-m downlink)! There's a message here.

When Phase 3C is launched, two new technosport activities will be introduced. The satellite-based FMT will add a challenging new dimension to the traditional HF versions since OSCAR moves with respect to both the signal source and the test participant. Doppler shifts will complicate measurements for the competitor.

An OSCAR-based F'n'H event is conceived as an analogue of the remarkable SARSAT/COSPAS search-and-rescue transponders on US and USSR satellites. In the AMSAT version, the "Fox" (QTH unknown) is a "cooperative" beast. To aid the hounds, the fox retransmits the WWV time standard through the transponder. By comparing the arrival time of the direct WWV signal (via the ionosphere) to the arrival time of the WWV signal, as relayed by the satellite, a curve is determined. A series of measurements narrows the unknown QTH to a region.

The FMT and F'n'H events are in the planning stages and will be described more fully in a subsequent column. Next time, we'll look at what it takes to be successful in the ZRO test.

[Write to AMSAT for *free* information about getting started with OSCARs. An SASE to AMSAT, PO Box 27, Washington DC, 20044 will do the trick.—Ed.]

Amateur Satellite Communications

Conducted By
Vern "Rip" Riportella, WA2LQQ
PO Box 177, Warwick, NY 10990

Fun, Games and (Hopefully) Education on OSCARs—Part 2

Last month, we discussed how AMSAT performs its receive sensitivity test, the ZRO Test. This month, we'll examine what it takes to do well in the competition.

There's no single piece of equipment or technique that will assure you of being among the top performers in the ZRO Test. Rather, it's a combination of factors that makes the difference. But if there's any common thread that unites the top performers, it's that they all have paid scrupulously careful attention to detail.

When striving for better receiving sensitivity, the primary obstacle is noise. The ability to hear anything is dependent on the power of the signal you're trying to hear compared to the power of the noise in the monitored bandwidth. This is most commonly called the in-band signal-to-noise ratio, or the S/N ratio.

The receiving system consists of two basic components: an antenna, which couples a weak electromagnetic field to your receiver, and a receiver, which converts the RF to AF and couples the resultant audio to your brain. Some important ancillary equipment consists of the preamplifier (actually an extension of your receiver), the transmission line connecting the antenna to the receiver, and the speaker (or earphones), which couples the receiver to your ears.

A preamp is essential, although where it's placed depends on several factors. In essence, the preamp should be placed ahead (on the antenna side) of any significant losses such as might be incurred in transmission lines. Why?

Picture your transmission line as a noise source (because it is). It's lossy and warm.[1] Any losses incurred between the antenna and the preamp contribute greatly to overall receiver S/N degradation. This is *the single most important thing you can do!* Get the preamp close to the antenna. How close? If there is a 1 dB loss between your antenna and the preamp, you're at the limit of what I consider a reasonable tradeoff. Keep in mind that at 2 m a 1 dB loss is incurred in about 40 feet of RG-8 foam cable (in new condition). At 70 cm, you incur 1 dB loss in about 25 feet of the same coax.

For weak-signal space work, you should have as much gain as possible generated by your antenna array. How much is enough? Start with a single Yagi. Point it to the zenith. Does the needle of your receiver's S meter come off the peg or does it sit there listlessly? If the receiver is in good shape and you have a good, low-noise preamp, the needle might be just off the peg. If not, try to borrow another preamp for a test. Put the second preamp in line. Now, the S meter should be well up-scale and your receiver's speaker should be erupting with the rushing sound of white noise. If the noise is not prominent, you may need a new receiver. If a single preamp gives you noticeable, but slight, needle movement, then chances are you have enough gain in the active components (preamp and receiver). Next you need to concentrate on the passive components, specifically the antenna array. You can add a second Yagi. If you separate it at the proper distance from the first antenna, you will reduce the beamwidth of the resulting array.

The sky is full of noise sources. When you add antenna gain, you are, in effect, increasing the S/N ratio in two ways. First, since the gain of the antenna goes "up," it couples more energy to the preamp. Thus, the ratio of the desired satellite signal energy from the antenna increases in proportion to the undesired noise energy unavoidably generated in the preamp. The S/N ratio is improved. Second, and somewhat more subtly, when the antenna gain is increased, the total sky area that is within the beam of the antenna decreases in direct proportion. So all those noise sources in space you don't care to hear are out of the beam. What you *do* want to hear, the satellite, stands out more clearly against the omnipresent noise background.

In general, therefore, once the receiver AGC begins to be "tickled" by the sum of sky noise plus preamp noise plus the noise of everything else ahead of the receiver, as indicated by very slight S-meter movement, it's time to increase antenna gain as far as practical.[2] More antenna gain is added until the target signal exceeds the total noise by an acceptable margin.

What about other noise sources? Noise from distant thunderstorms, the bane of DX HF work, is virtually nonexistent at VHF and above because the ionosphere rarely reflects VHF signals. Thunderstorm noise simply passes out into space for the most part. Automobile ignition noise and noise from urban industrial centers is less of a problem with VHF/UHF space communications because the antennas are generally pointed skyward. You thus turn a "deaf ear" toward the terrestrial noise sources.

But, there are noise sources much closer to home that can degrade your station's performance. Your azimuth and elevation rotators probably contain noise-generating electric motors. Slewing these motors probably increases your local noise. Computers and related digital equipment (such as packet radio TNCs) can generate noise into the GHz range. Turn them off when possible. If not, place them in a thoroughly RFI-proofed box. Turn off all TVs and microwave ovens.

Don't use BNC connectors if you can avoid them, and never use them outdoors. The bayonet connections are notoriously noisy when corroded. Use type "N," TNC or SMA connectors instead. Tighten all antenna hardware. In the most modern designs, the elements are insulated from the boom to avoid the intermittent contact and noise that results from inevitable corrosion of parts exposed to the weather. Make certain all station grounds and bonding straps are clean and tight. A poor ground can doom your otherwise excellent station to virtual deafness.

Is your receiver itself a significant noise source? If it uses a synthesizer as a frequency source, it could be. So-called phase noise can be a problem with some VHF/UHF radios just as this has recently become recognized as a factor in the noise floor of HF radios.[3]

Another way to improve the S/N ratio is to reduce receiver bandwidth. This can mean merely using the narrowest CW filter available commensurate with the signal spectrum of interest. In the ZRO Test, 10-WPM CW is sent so even the 200-Hz filters some top-notch radios offer are practical. Still, more improvement can be obtained by using advanced variable bandwidth audio filters. Some ZRO Test participants, using audio filtering, have successfully copied CW signals others couldn't even begin to detect.[4]

What is in use at the top stations? On 2 m, the top stations are using antennas with gains of 18 or 19 dBi, and a preamp with a noise figure in the 1 dB (or less) range and gain of 18 to 20 dB. A median good station generally runs 12 dB antenna gain and approximately the same preamp as the top station. For 70 cm, the top stations use antennas providing 20 or 22 dBi gain with a preamp in the 0.6 dB noise figure range, while yielding gain in the 18 dB range. A median good station on 70 cm runs about 13 or 14 dBi gain on the antenna and about the same preamp as the top stations. In all cases the preamp is at the antenna.

Especially at 70 cm and up, the sky is generally quiet and even a relatively weak signal from a satellite clearly stands out against the cold sky. For example, it is not uncommon for a first-rate ZRO competitor to be able to hear a milliwatt-level 70-cm downlink signal at a range of 22,000 miles!

At the low end of the capabilities scale, many regular satellite users get by with just 10 dB of antenna gain and 4-5 dB noise figure on both 2 m and 70 cm. They can hardly begin to hear the weak QSOs. The ZRO Test demonstrates how much room for improvement there is at some stations.

When we recognize how easy it is to have a station that hears well, it seems strange indeed that some would hog-tie themselves with ineffective or no measures at all to hear all that's on the satellites. The ZRO Test seeks to provide a means of evaluating both objectively (in absolute terms) and relatively (compared to one's peers) how well one's satellite station really hears. In that sense it has been an outstanding success.

Not only will the ZRO Test continue as soon as possible on AO-10 and Phase 3C, but, as mentioned last month, the ZRO Test is destined to foster a whole new breed of on-satellite competitions in the tradition of techno-sport. Next time we'll look forward to the new breed of techno-sport activities planned for the future.

Notes

[1] Any device that is above absolute zero degrees (−273°C) generates noise that is attributable to thermal agitation of its molecules and other lesser effects.

[2] When your receiver is sufficiently sensitive and quiet, such that sky noise comprises the majority of the noise to which it is subject, the receive system is said to be "sky noise limited." This is the point of departure for further improvements obtained through increased antenna gain.

[3] G. F. McCanless, Jr, KA4GSQ, "Using QSTs to Choose an Old HF Rig," QST, Feb 1987, p 20.

[4] Jeff Bishop, W7ID, holds the world record. He is the only one ever to earn a Z8 rating on the Mode B (2-m downlink) session of the ZRO Test. He used special home-brew antennas, but we suspect his real advantage was in the very narrow (about 40 Hz) audio filter and earphones he used.

Information about satellites and AMSAT can be obtained for an SASE to AMSAT NA, PO Box 27, Washington DC 20044.

Amateur Satellite Communications

Conducted By
Vern "Rip" Riportella, WA2LQQ
PO Box 177, Warwick, NY 10990

Fun, Games and (Hopefully) Education on OSCARs, Part 3

Last month I explained what was required to excel in AMSAT's receive sensitivity test, the ZRO Test. This month let's pick up a new direction within the general topic of satellite techno-sport.

Amateur Radio has, at its heart, the notion of public service. Indeed, in the public consciousness, the image of amateurs rallying to provide vital communications during emergencies is one we actively promote. And in no small measure is that image unwarranted. In the future, Amateur Radio satellites could play a significant part in emergency communications in any of several ways. AMSAT's Phase 4 geosynchronous satellite plans include a significant public service component at their heart.[1]

There are possibilities and opportunities for experiments leading to real life-saving techniques that can commence now using current satellites. To dramatize this, to gain experience and proficiency in the field and to promote productive use of space resources, AMSAT is considering broadening its techno-sport base well beyond just the ZRO Test I've addressed over the past two months. In the future, satellite users can look forward to a fascinating new type of techno-sport that is both fun and fairly easy, but has some important implications for the whole field. I am talking about radiolocation by satellite.

In a techno-sport context, radiolocation by satellite is simply the game of trying to locate a hidden transmitter by using its satellite uplink signal together with special information about the satellite. In a real-world emergency, this technique can be used to locate downed aircraft or other vehicles in distress. In 1975, AMSAT pioneered a similar concept using the AMSAT-OSCAR 7 spacecraft. This led to the development of NASA's Search and Rescue Satellite (SARSAT) system.[2,3]

SARSAT made its first save after locating a downed Canadian aviator on September 10, 1982. Since then, SARSAT and its Russian counterpart, COSPAS, have together saved over 500 lives, mostly in remote territory. The principles are simple and, as will be shown, can be easily and reliably transported to Fuji-OSCAR 12, the newest OSCAR, which was launched in 1986.

Most aircraft and many ships have emergency locator transmitters (ELT) aboard. In an emergency the ELT is automatically activated. When its signal is detected by a SARSAT/COSPAS-equipped satellite, the ELT's location can be determined almost immediately, depending on the precise circumstances. Rescue units can then be dispatched quickly to the scene. The essential elements to this scenario are these: The lost vehicle has a radio aboard, but it is out of range of terrestrial VHF or UHF direction-finding (DF) equipment. But the SARSAT/COSPAS satellites, listening on the ELT frequencies of 121.5, 243 or 406 MHz, easily pick up the ELT transmissions and relay them to ground analysis sites. Here the Doppler shift of the ELT signals heard by the SARSAT/COSPAS satellites is analyzed, and a position is computed for the vehicle in distress.

The ELT's emissions can be detected anytime an ELT is within the satellite's footprint, which is about 4000 miles in diameter. Once an ELT emission is detected, the signal is transponded (relayed) to a Local User Terminal (LUT) in real time for analysis and computation of the ELT's QTH. Using the older version's system, the relay takes place in real time, requiring both the ELT and the LUT to be in the footprint concurrently. On newer versions, the satellite makes a time/frequency record of the ELT and dumps the record to the LUT on command. The LUT can thus obtain Doppler data on ELTs that the satellite has "seen" previously in its travels. Moreover, on the newer systems, the potential ambiguities inherent in the data provided by the older versions are quickly resolved within a few minutes from a single satellite.[4]

The SARSAT/COSPAS satellite contains a simple transponder that listens on 121.5, 243 or 406 MHz and regenerates the downlink at 1544.5 MHz. The analysis sites, established by the US, USSR, Canada and France, subject the resulting Doppler shift to careful scrutiny. The result is a quick determination of the ELT QTH, even in extremely rugged terrain.

Doppler shift is related to the closing or diverging velocity of the satellite to or from the ELT. Fig 1 shows the general shape of the curve of frequency versus time resulting from a SARSAT/COSPAS satellite approaching and then diverging from an ELT. The SARSAT/COSPAS ground station would receive a set of frequency and time data such as that which produced this curve. At acquisition of signal (AOS), the Doppler shift is maximum and positive; the signal heard on 1544.5 MHz is upshifted.[5] The Doppler shift decreases until the satellite reaches its closest point to the ELT at the Time of Closest Approach (TCA). At this point, the satellite moves in a tangential line to the ELT. Thus, the satellite is neither converging nor diverging from the ELT and the observed frequency is the actual ELT emission frequency, F0. Then, the satellite begins to move away from the ELT and the Doppler effect causes the observed frequency to be below the actual frequency as shown in the figure. Finally, at Loss of Signal (LOS), the Doppler shift has caused the observed frequency to be at its lowest.

The mathematics underlying the ELT QTH determination are fairly simple. Knowing the ELT frequency, the observed frequency and the position of the satellite at various times is sufficient. The angle between the path of the satellite and the ELT at any instant is found by calculating the arc-cosine of the ratio of observed frequency to ELT frequency multiplied by the ratio of the speed of light to the satellite's velocity.[6]

Using the same principles, SARSAT/COSPAS radio-location techniques can be performed on FO-12. FO-12 has an uplink on 2 m and a downlink on 70 cm. Using the Mode-JA transponder, it should be possible for a clever, well-equipped amateur to locate a hidden transmitter to within a small region in just a few minutes.

To test this theory, AMSAT will begin a series of demonstrations on how the technique works in the near future. Simultaneously, AMSAT will publish a guide to calculations and required instrumentation. At a minimum, the equipment required will include a precision frequency counter to measure the downlink frequency, a digital clock and a computer to display the position of the satellite and to provide a detailed ephemeris of its time and position.

To make the demonstrations more interesting, AMSAT will format the event in the guise of a techno-sport competition. Certificates will be offered and the most accurate participants will be specifically cited for their achievements.

Next month we'll investigate yet another radio location by satellite technique using the so-called time-difference of arrival (TDOA) principle. For more information about AMSAT and OSCAR satellites, send a business-sized SASE to AMSAT, PO Box 27, Washington, DC 20044.

---

[1]J. King, V. Riportella and R. Wallio, "OSCAR at 25: The Amateur Space Program Comes of Age," *QST*, Dec 1986 and Jan, 1987.
[2]*AMSAT Satellite Report*, No. 43/44, Oct 11, 1982.
[3]*AMSAT Newsletter*, Vol VIII, No. 1, Mar 1976.
[4]A more detailed system analysis will appear in a future edition of *QEX*, the advanced experimenter magazine from ARRL and AMSAT.
[5]Assuming the ELT is on 406 MHz and the 1544.5-MHz downlink Doppler shift is subtracted out, the net Doppler shift for the ELT, as heard from SARSAT/COSPAS, is about +10 kHz, maximum. Satellite velocity in orbit is about 4.5 miles per second.
[6]The *QEX* article will discuss potential ambiguities and how in practice they are eliminated or mitigated.

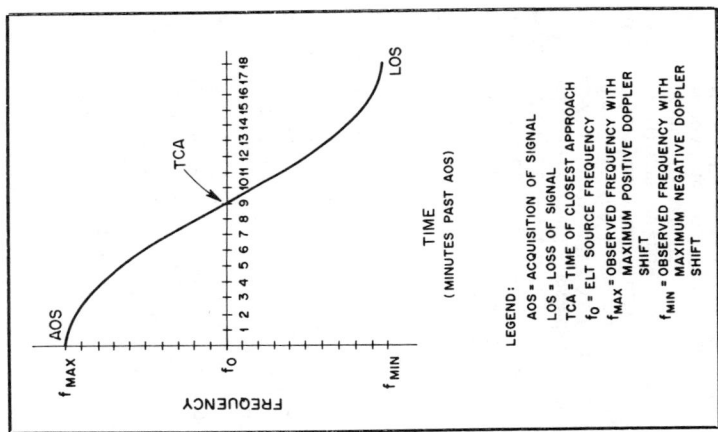

Fig 1—ELT typical Doppler-shift curve as observed by a typical SARSAT.

Amateur Satellite Communications

Conducted By
Vern "Rip" Riportella, WA2LQQ
PO Box 177, Warwick, NY 10990

Fun, Games and (Hopefully) Education on OSCARs, Part 4

Last month, I introduced the idea of hidden-transmitter location using OSCARs and the SARSAT/COSPAS series of international satellites. This month, I'll continue with the theme of locating a hidden transmitter using a satellite. As before, this discussion is within the context of Techno-Sport, an AMSAT concept for learning interesting and useful tools and methods in radio and space technology while having fun in the process.

Measuring the Doppler shift of an emitter (such as Fuji-OSCAR 12) or a reflecting object (such as your automobile when illuminated by the local constable's radar) is a convenient and easy means of detecting relative motion. The problem is that when the relative motion is low (in terms of wavelength and velocity), Doppler shift is low, too. Thus, with police radars, there is a lower limit to the relative velocity that can be measured. Similarly, if the frequency of a satellite source is low enough or its relative motion is sufficiently low, the Doppler shift may be too low to measure with any useful degree of accuracy.

For example, when AMSAT-OSCAR 10 was first launched, one of the first orders of business was to locate it precisely. This was not as easy as it may seem, because, even though the Air Force's SPACETRACK system can and does track thousands of objects the size of AO-10 at geosynchronous altitudes,¹ the most sensitive tracking systems are reserved for important defense missions. The more common, lower-priority tracking systems are hard-pressed to track an object the size of AO-10 several tens of thousands of miles distant. So, AMSAT engineers devised their own means of tracking AO-10 that did not rely on radar, per se. Because AO-10 moves relatively slowly (especially at apogee, the highest point in its orbit), the Doppler shift method is not suitable. Instead, AMSAT engineers measured the round-trip transit time for pulses originated on the ground.

This system is simple in concept and, except for the sophisticated number-crunching techniques applied to the resultant data, simple in practice. A pulse is sent to AO-10 on its 70-cm uplink. Simultaneously, a clock is started. The AO-10 Mode B transponder receives the pulse and regenerates it on the 2-meter downlink. The returned pulse is received by the originating station and the clock is stopped. Because the transponder delay time is known (from prior measurements made on the ground), and the propagation velocity of the radio wave is known, it is easy to convert the measured round-trip transit time into a range (distance) from the earth station to the satellite at that instant. The result of this measurement does not determine the position of the satellite, merely a range. To be slightly more precise, the satellite has been determined to lie on the surface of a sphere, the radius of which is the measured range.

To isolate the satellite's position to a relatively small domain on the surface of the sphere requires more measurements. One approach is to make additional measurements from other locations. Thus, a series of incongruent spherical surfaces is determined. Ideally, they all will intersect at a point—the location of the satellite at that instant. A real-time computer analysis could, in theory, provide real-time tracking this way. In practice, however, the measurements are fed into a computer program and the orbital characteristics themselves are determined.² Given this, the position of the satellite can be predicted at any time in the future, because the motion is fully understood in time and space.

The uncertainties in measuring range to the satellite and reducing range to location are many and varied. Moreover, because of irregularities in the data and miscellaneous uncertainties, what normally results is a series of onion-skin-like probability densities. The outer shell represents high probability that the satellite is contained within the volume of that shell. That is, the more precisely one says he knows the location, the more likely he is to be wrong. Conversely, it's easy to say the satellite lies within a 1000-mile sphere, but quite difficult to say it's within a 10-mile sphere. As more data is added and the measurements are refined, the volume of space representing an arbitrary uncertainty value decreases. If the process of data collection and analysis were to continue indefinitely, the onion would shrink to a point of 100% certainty.

Using these and similar methods, AMSAT engineers were able to locate AO-10 to within a few cubic miles over a range of about 25,000 miles.³ A more advanced version of this technique may be employed in the future.

What would happen if we turned the situation around a bit? In the foregoing example, the locations of the tracking stations were well known, but the orbit was initially unknown. Suppose the opposite were the case. Let's say the orbit is well known, but your location is unknown. Could you determine your own location? The answer is definitely yes. In fact, a vast array of satellites (called Global Positioning Satellites, GPS) is in use by the government and private sectors for precisely that purpose. With GPS, several satellites are in view of your location at any given time. You have a special receiver with a powerful portable computer built in. The GPS send precise time signals. The computer in your receiver measures the arrival time of the signals. By comparing the arrival time of the signals from several satellites, whose precise orbits and positions the computer knows, the computer can work backwards to determine *your* position to within a few dozen feet.⁴

By rearranging the knowns and unknowns a bit, one can structure an interesting Techno-Sport challenge for those OSCAR enthusiasts who like to test their technical talent in an interesting and educational event. By adding some accuracy incentives, a healthy competitive event results. This is how it might be structured.

An "unknown uplinker" sends a signal on the uplink of AO-10 (or Phase 3C). You're listening on the downlink. You know exactly where AO-10 is (from your AMSAT orbital data and your computer), thus you know how long it took for the downlink signal to reach you.⁵ But, you don't know how long it took the uplinker's signal to reach the satellite, so you can't locate him.

Now suppose the signal is one that carries special timing information. Let's say the uplinker takes a WWV signal he receives on HF and retransmits it on the uplink. You listen to WWV on one receiver, and the downlink on the other receiver. The difference in arrival times of the signal direct from WWV and the signal as relayed by the uplinker through the satellite contains most of the information you need to determine the uplinker's location.

The other piece of information needed is your distance from WWV. More precisely, you need to know the radio path distance from your location to WWV. Because you probably hear WWV via skywave, the exact path length may be difficult to ascertain. This adds one of the major uncertainties to the determination of the uplinker's location. But, with reasonable care in measurement, an estimate of the location should be possible in an hour or less of measurements. Moreover, by combining your measurements with someone else's at a widely disparate location in a cooperative determination, the uncertainties involved should be reduced.

The solution to the problem lies in establishing the distance from the uplinker's location to the satellite, because all of the other distances are known.⁶ When you hear the WWV signal relayed by the satellite, you can determine how far away from the satellite the originator is. Because the satellite is moving, if but slowly, the transit time between the satellite and the uplinker changes, too. By carefully logging the changes in path length, the uplinker can be located by using a succession of intersecting spheres as described earlier. With time, the uncertainty in the exact location is reduced until a good guess can be made as to where the "unknown uplinker" is.

I'll be off until September. When I return, I'll discuss in more detail the workings of this newest technique in radiolocation by satellite in the Techno-Sport context. To get more information on this and other AMSAT satellite projects, write to AMSAT at the address given below.⁷

Notes

¹About 22,300 miles.

²An ephemeris is generated that provides a mathematical means for describing the satellite's position along with the time it will appear at that position. The Keplerian data provided by NASA via AMSAT is a special form of this data.

³That is, a few cubic miles out of a potential volume of over 60 trillion cubic miles of space!

⁴Several automobile manufacturers have demonstrated GPS systems in prototype cars. A receiver, computer and special map projection on a CRT show your exact location even while you're moving. Other companies are mounting devices on their vehicles that will report their position to headquarters so each vehicle can be closely monitored.

⁵Radio waves travel at about 186,300 m/s in a vacuum. That works out to about 1 foot per nanosecond, or about 1017 feet per microsecond. Measuring a microsecond in your typical ham shack is a lot easier than it formerly was, thanks to inexpensive digital counters.

⁶Your location, the satellite's location, WWV location and all the distances between them can be determined.

⁷Information about AMSAT can be obtained by sending a business-size SASE to AMSAT, PO Box 27, Washington, DC 20044.

Amateur Satellite Communications

Conducted By
Vern "Rip" Riportella, WA2LQQ
PO Box 177, Warwick, NY 10990

Prospects for Mobile and Portable OSCAR Operation

Never is the advantage of using OSCAR satellites more obvious than when the sunspot cycle is at its low ebb. While the MUF hovers in the 15-MHz range, and would-be 10- and 15-meter band users reluctantly slide into the ever-more-crowded 20-meter band, satellite use goes merrily on its way!

OSCARs historically have used frequencies that were only mildly affected by F2-layer density. Mode A transponders, such as those used on AMSAT-OSCARs 6, 7 and 8 and all the Russian RS-series, use a 2-meter uplink and a 10-meter downlink. Occasionally, during periods of high solar activity, the ionosphere becomes sufficiently radio-opaque to noticeably attenuate 10-meter radiation through the ionosphere. Listening to AO-7 or RS-1 as it passed "behind" various attenuating zones while its apparent signal strength varied by 10 dB or more was an educating experience. It was as if a buddy of yours were groping through the fog with a flashlight. One moment he could be seen fairly clearly, the next only dimly.

There are, however, more reasons to want to move up in OSCAR frequency than just to avoid the occasional highly attenuating ionosphere. Spectrum capacity and obtainable antenna directivity are two other very strong motivations for moving to the UHF regime in future satellites. In fact, the current AO-10 satellite has set the course with its combination Mode B and Mode L transponders.[1] Future satellites such as Phase 3C, to be launched later this year, will use a combination of Modes B, J, L and S (see Table 1).

Never is it as obvious why it is strongly desirable to move up in frequency with OSCAR than when one closely looks at future satellite systems. Many future satellites will continue the high-orbit tradition begun with AO-10. In its high, elliptical orbit, AO-10 can relay signals originating from nearly a hemisphere, putting in contact all those within its view. And it can do this for hours on end. Phase 3C will continue this tradition in another elliptical orbit with improved characteristics for the higher latitudes. Discussions have begun regarding orbital options for a possible follow-on called Phase 3D. Some very clever combinations of two or more satellites in true Molniya (elliptical) orbits yield many of the desirable characteristics of a geosynchronous orbiting satellite while reducing substantially the cost and engineering burden of the latter. In any case, the future high-flying satellites will glimpse much of the earth during the course of their orbit.

Thus, while viewing so much of the earth's surface, it is necessary to have a QSO capacity (spectrum) commensurate with the potential user-load. Where can the necessary spectrum of several hundred contiguous kilohertz be found? Quite plainly, the only option is in the UHF regime.

Another reason for the appeal of UHF use with OSCAR is the reduced size of highly directional antennas as compared with VHF or HF antennas of comparable gain. Using large, heavy antennas on a spacecraft is not becoming available, it is conceivable to have highly portable or even mobile-in-motion OSCAR QSOs on a regular basis in the not-too-distant future, say 3 to 4 years. The gating technologies are just about available, and early conceptual designs for both Phase 3D and Phase 4 (geosynchronous) systems suggest a strong potential for portable and/or mobile use. But the concept rests totally on using UHF and SHF frequencies to achieve the link budgets required for reliable communication.

With just a smidgen of imagination one can envision satellite-portable terminals being erected at remote locations for emergency communications or even sport. Imagine pulling into a major disaster zone with your four-by-four truck, unloading a few modest shipping containers, assembling a half dozen connectors, setting up a tripod, aiming at the sky and being in touch with the hemisphere within a half hour of arriving! Further, imagine connecting a portable FM repeater to your satellite terminal making it a gateway terminal and affording local 2-meter emergency communicators a channel to the satellite and hundreds of other stations as required. Or, by interfacing a portable linear translator to your terminal, provide integrated digital and voice services in/out of the affected area. Here we might find high-speed packet data working side by side with voice QSOs linking the affected area with support activities across the hemisphere.

Perhaps the ultimate in OSCAR technology will be attained when OSCAR mobile-in-motion QSOs are regularly afforded. Yet the day when this too will be reality is fast approaching. The enabling technology may be available as derivatives of NASA's so-called MSAT-X project. Later this year the FCC is expected to rule on several commercial applicants' proposals to implement a mobile satellite communications system. It is a distinct possibility that amateurs may obtain this capability nearly concurrent with commercial applications.[2]

Whether for spectrum availability, ionospheric transparency or reducing the size of antennas, the move to higher satellite frequencies is both desirable and inevitable. The prospect of promoting smaller, more efficient earth stations for use by mobile, portable and apartment-dwelling amateurs is a highly attractive one. And the prospect of using gateway access and integrated digital/voice channels over future OSCARs builds on concepts now being architected.[3]

Next month we'll take a closer look at what a gateway is and how it functions.

Notes

[1] Mode B uses a 70-cm uplink and a 2-meter downlink. Mode L uses 24 cm up and 70 cm down.
[2] Interestingly, one of the commercial proposers for MSAT-X would use technology first used aboard AO-10. Thus, in this sense, Amateur Radio satellites are carrying on a venerable tradition of ham radio dating back decades, ie, technologies developed by amateurs finding commercial applications.
[3] Information on getting started on OSCAR may be obtained from the author for a business-sized SASE to the address shown above.

Table 1
Phase 3C Transponder Frequencies

Mode B
Uplink: 435.425-435.575 MHz
Downlink: 145.975-145.825 MHz
General Beacon: 145.8125 MHz
Engineering Beacon: 145.975 MHz

Mode JL
L Uplink: 1269.575-1269.325 MHz
L Downlink: 435.725-435.975 MHz
General Beacon: 435.650 MHz
J Uplink: 145.82-145.86 MHz } Option 1
J Downlink: 435.93-435.97 MHz }
or
J Uplink: 144.44-144.48 MHz } Option 2
J Downlink: 435.93-435.97 MHz }

Digital Mode L (RUDAK)
Uplink: 1269.675 MHz
Downlink: 435.675 MHz

Mode S
Uplink: 435.625 MHz
Downlink: 2401.337 MHz
S Beacon: 2401.267 MHz

Notes

(i) It is currently undecided which of the two J input options shall be implemented. In either option, there is an overlap whereupon signals originated at 24 cm and 2 meters can result in a downlink between 435.93 and 435.97 MHz.
(ii) RUDAK is a digital-only transponder that will use PSK.
(iii) Mode S will be a soft-limited FM transponder that can also accommodate up to four SSB signals.
(iv) Although minor revisions in frequency may be anticipated, the frequencies presented are those currently planned.

---

just inconvenient, it's impossible. Launcher constraints of size, mass and cost limit the types of antennas to be used aboard OSCAR. From a high-flying OSCAR such as AO-10, the earth subtends an angle of about 16 degrees when viewed from apogee. So it's notably wasteful to scatter RF across the cosmos using an antenna with, say, a 45-degree 3-dB beamwidth. Rather, one wants to concentrate its RF toward the earth. Given the constraints mentioned, one can best achieve this using UHF or above and the compact, highly directional antennas achievable there.

A similar situation exists with your ground station. You want to concentrate as much power as possible in the direction of the spacecraft and, conversely, avoid spraying the sky with RF that misses the target. Again, this is most effectively achieved with a given-sized array when the wavelength is relatively small, ie, the frequency is in the UHF (or higher) regime. "Well," you might retort, "I've got a few acres of real estate here and don't mind putting up a real monster antenna to work OSCAR." But even though you might thoroughly enjoy merely the prospect of establishing an aluminum monster to work future OSCARs, and even if we ignored the strong spacecraft engineering motivations for using UHF and above, using VHF and below would make much less likely one of the most striking and powerful future prospects of OSCARs: portable and/or mobile OSCAR use. Using UHF and SHF equipment now

Amateur Satellite Communications

Conducted By
Vern "Rip" Riportella, WA2LQQ
PO Box 177, Warwick, NY 10990

Waves in Rotation: The Challenge of Circular Polarization

Few topics in Amateur Radio elicit the intense fascination of antennas. But the "field" really becomes interesting when antennas designed for space communication are factored in.

In space communications, the earth's magnetic field can dramatically affect radio waves traveling over long distances. The effect is to rotate the plane of polarization. To reduce the adverse effects of this polarization rotation, antennas designed for space communication often employ a special form of polarization called "circular polarization" (CP), a combination of the familiar vertical and horizontal polarization generated by common antennas. What makes CP desirable for space communications?

It was known early in the 19th Century that some substances are optically active; the plane of polarization of light passing through these substances is rotated. In 1845, Michael Faraday discovered that some substances, which normally show no optical activity, can be made to rotate polarization by subjecting it to a strong magnetic field. This came to be known as the "Faraday Effect."

Radio waves and light waves are basically the same phenomenon. Early radio experimenters soon found that the Faraday Effect was detectable in low-frequency radio waves where the geomagnetic field works to rotate polarization. The amount of rotation of the wave's plane of polarization depends on its frequency, how far the wave travels in the magnetic field, how strong the field is, and the wave's angular relation to the magnetic field lines. Briefly, the degree of wave plane rotation *increases* with:

1) increasing magnetic field strength
2) increasing path length
3) increasing alignment of the wave plane and field lines

The degree of wave plane rotation *decreases* with increasing frequency, ie, VHF is affected less than HF, UHF less than VHF, SHF less than UHF, and so on.

To someone watching TV Channel 7 at a range of 50 miles, the Faraday Effect is negligible. But take roughly the same frequencies (around 200 MHz) and stretch the path over 10,000 or 20,000 miles, and Faraday Effect rotation (or simply Faraday Rotation, FR) can greatly affect received signal polarization.

The problem arises when, owing to random incoming wave polarizations, the wave and the receive antenna are not in the same plane. The result is a widely varying received signal strength because of cross-polarization losses. Cross-polarization losses are familiar to anyone who has had difficulty using a horizontal Yagi to access a vertically polarized repeater antenna. An orbiting spacecraft transmitting to earth over a long path using UHF could very well originate a signal using horizontal polarization only to have it arrive vertically polarized. Frequently, the case is that the arriving wave's polarization is essentially random and changing at any instant. The solution most often used for reliable space communications is two-fold. Since FR is less of a problem with increasing frequency, space communications are best accomplished at the highest practical frequency. And, when FR at the chosen frequency can still be a problem, CP is used. In a practical sense, the frequencies between 0.1 to around 2.0 GHz is the range where CP finds its greatest utility. Frequencies below 0.1 GHz are seldom used for space communications; frequencies above 2.0 GHz are affected by FR only minimally. But, since many of the OSCAR users' prime frequencies lie in the area where FR is significant (0.1 to 2.0 GHz), CP antennas must be employed for best results.

The CP antenna puts equal amounts of RF in each plane. The result is that the effects of FR are nullified. The price paid, however, is that the power in any given plane is less than if linear polarization were used. If one knew what the polarization would be all the time, one would use linear polarization and ensure the receive antenna was well aligned with it. But since the factors that control FR may all be varying at once, putting power into all planes with CP assures that when the signal is received on a CP antenna, the *average* received signal power will be higher than if linear polarization had been used.

Moreover, with OSCAR satellites in particular, the additional QSB caused by the spinning of the spacecraft most often makes CP doubly desirable. Indeed, many antennas built into OSCAR satellites are specifically designed for CP to reduce both FR and spin caused by QSB.[1]

CP is somewhat more difficult to generate than linear polarization. A CP antenna combines vertical and horizontal components in a particular time relation or "phase." This is accomplished most easily using an antenna not encountered in HF work, but quite familiar in the space communications context: the helix.

The corkscrew-like radiating element of the helix radiates axially a wave whose plane of polarization rotates as it leaves the transmitting antenna. Imagine a ribbon twisting off into the distance as the wave travels away from the source helix. The appearance of the twisting ribbon gives us a clue as to why CP works well in space communications. If the receive antenna is linear, no matter what its current orientation,

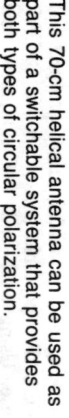

This 70-cm helical antenna can be used as part of a switchable system that provides both types of circular polarization.

the wave will never be cross polarized for very long. If the receive antenna is CP, the wave is always in the "right" plane as it's received. In other words, a CP antenna "hears" equally well in all planes.

The direction the ribbon twists is called the "sense" of the wave. Right Hand CP (RHCP) rotates clockwise as seen from behind the source antenna, and Left Hand CP (LHCP) rotates counterclockwise. Thus, the twist of the imaginary ribbon is either clockwise or counter-clockwise, corresponding to RHCP or LHCP, respectively. The rate of twist is one complete turn (360 degrees) per wavelength. For best results, both the transmitting and receiving antennas should have the same sense.

Helices may be the easiest way to generate CP and one of the most popular types of home-brewed satellite antennas. It's not the only way to get CP, by far. Many OSCAR users employ a pair of specially phased Yagis to generate CP in the 2-m through 24-cm bands. The two identical Yagis have their elements mounted orthogonally (at right angles; thus, "crossed Yagis"), either in two separate antennas side by side or two separate Yagis sharing a common boom.[2]

CP can be generated by a variety of antenna types. The use of CP antennas for space communications is essential in many cases. Moreover, the bizarre shapes and configurations used to generate CP serve to enhance the already entrenched mystique that antennas enjoy with amateurs.[2]

---

[1] The QSB caused by the spinning of the satellite is called "spin modulation." On frequency, it sounds like rapid amplitude modulation, or flutter, of the downlink, and depends on the angular velocity of the spacecraft about its spin axis and the specific antenna pattern involved.

[2] To learn more about some of the interesting aspects of Amateur Radio space communications, write to AMSAT for free information about getting started with OSCARs. An SASE to AMSAT, PO Box 27, Washington, DC 20044 will do the trick.

Amateur Satellite Communications

Conducted By
Vern "Rip" Riportella, WA2LQQ
PO Box 177, Warwick, NY 10990

Where to Get OSCAR Information

In recent months, we've been expanding our satellite knowledge in specific areas. We've learned the basics of satellite operating and how to establish a station. Most recently, we've examined some of the activities that occur on the OSCARs. I introduced the theme of Techno-Sport to denote a fun-type activity that has a strong learning component involved. For example, in the ZRO-Receive Sensitivity Test, I explained how participants in this Techno-Sport "contest" garnered awards for superior station performance. They proved they could hear better than most other stations in a realistic on-the-air test. Later, I introduced the Radio-Location Techno-Sport, an activity planned to commence later this year. Additional details on this exciting new aspect of Techno-Sport, which has obvious parallels in the COSPAS-SARSAT search-and-rescue satellite area, will follow in this column.

I thought it appropriate to provide an "information handle" in this month's column, since it's my experience that the hardest part of getting started in any new activity is knowing *where* the appropriate information spigots are, and how to turn them on. Once they're turned on, we can ask the right questions to get the information we need to get on the air. So, here are some suggestions of places to find information on operating OSCARs.

First, and perhaps easiest to access at no cost whatsoever, are the on-the-air nets (see Table 1) that AMSAT and its overseas partners sponsor. Some of these nets are over a decade old; others are new-starts. In any case, you'll find them a treasure trove of current information and helpful hints. Most net-control operators are quite knowledgeable and willingly reply to questions from the net.

An extensive list of packet-radio bulletin boards, numbering several hundred around the world, carry AMSAT News Service bulletins. These bulletins are carried by many voice-net stations as well, but far more packet BBSs carry the bulletins. The BBS list is too long to publish here, but is available from me for a business-size SASE.

AMSAT Area Coordinators are an excellent source of information, too. There are more than 120 Coordinators across North America, and most are glad to help get you going. Perhaps one near you can provide just the helpful hint to break that OSCAR QSO logjam you've encountered. You can send me a business-size SASE for a list of Area Coordinators.

Written OSCAR information abounds—if you know where to look. AMSAT North America (AMSAT-NA) is the largest of nearly two-dozen affiliated AMSAT organizations around the world; AMSAT-NA publishes several periodicals. Its newsletter, *Amateur Satellite Report*, is a member service published biweekly. *QEX* is a monthly publication offered jointly by ARRL and AMSAT, and designed to appeal to the more technically inclined amateur. More and more satellite and advanced technology articles are appearing in *QEX*.

The Satellite Experimenter's Handbook published by ARRL is clearly the best all-around book in its field. Its comprehensive and authoritative approach make it required reading for the serious student of OSCAR and weather satellite work.

The most advanced topics are covered in AMSAT's *Technical Journal*. Here are presented professional engineering level papers accessible to advanced amateurs. *ATJ* is aperiodic and available from AMSAT HQ.

Helpful guides are provided in AMSAT's Beginner's Manual and the Phase III Operations Manual. The former takes novitiates from ground zero through their first OSCAR contact. The latter provides a thorough how-to-do-it for working the high-flying OSCARs such as AMSAT OSCAR 10 and the soon-to-be-launched Phase IIIC.

Just as it takes a good recipe to bake a good cake, it takes the right information to work OSCAR easily and consistently. Make use of the aforementioned information sources, and soon you'll be up there with the rest of the proficient OSCAR satellite users.

Information about AMSAT can be obtained by sending a business-size SASE to: AMSAT, PO Box 27, Washington, DC 20044. *The Satellite Experimenter's Handbook* and *QEX* are available directly from ARRL. All the publications mentioned are available from AMSAT.

Table 1
AMSAT Information Services Worldwide
(Updated as of June 1, 1987)

Service Area	Day	Time	Freq (MHz)	NCS (Primary)	Notes
International					
International	Sunday	1900 UTC	14.282	WD0HHU	1
International	Sunday	1900 UTC	21.280	WD0HHU	2
South Pacific	Saturday	2200 UTC	14.282	W6SP	
South Pacific	Saturday	2230 UTC	21.280	W6SP	13
Southern, Central and					
Eastern Africa	Sunday	0900 UTC	14.280	ZS6AKV	
"	Sunday	0900 UTC	7.080	ZS6AKV	
"	Sunday	0900 UTC	3.718	ZS6AKV	
"	Sunday	0900 UTC	3.665(AM)	ZS6AKV	
National					
Australia	Sunday	1000 UTC	3.685	VK5AGR	8
England	Sunday	1015 local	3.780	G0AUK	
England	Mon + Wed	1900 local	3.780	G0AUK	
Sweden	Sunday	1000 local	3.740	SK4TX	
Regional					
US East Coast	Tuesday	2000 local	3.840	WA2LQQ	3
US Central	Tuesday	2100 local	3.840	W0CY	3
US West Coast	Tuesday	2000 local	3.840	N6TE	3
Sub-Regional and Local					
England/Brighton Area	Sundays	1915 local	144.280	G6ZRU	
Scotland/Paisley	Daily	0900 local	144.625	GM1SXX	9
South Africa/J'Burg	Saturday	0900 UTC	145.650	ZS6AKV	
South Africa/J'Burg	Thursday	1830 UTC	145.650	ZS6AKV	
South Africa/Cape Town	Thursday	1730 UTC	145.750	ZR1KE	
South Africa/Durban	Thursday	1730 UTC	145.650	ZR5JJ	
South Africa/Pieter	Thursday	1730 UTC	145.750	ZR5JJ	10
South Africa/Pretoria	Thursday	1830 UTC	145.775		
South Africa/Pretoria	Thursday	1830 UTC	3.718		
South Africa/Pretoria	Thursday	1830 UTC	3.665		
South Africa/Port Eliz	Thursday	1830 UTC	145.775	ZR2FK	
US					
CA Los Angeles	Wednesday	2000 local	144.144	W6SP	
CA Los Angeles	Daily	0730 local	144.144	W6KAG	5
CA Los Angeles	Saturday	2200 UTC	144.144	W6SP	4
CA Los Altos	Tuesday	2000 local	147.150	WB6GFJ	
CA San Diego	Wednesday	1930 local	145.660	WB6LLO	
CO Denver	Wednesday	2000 local	147.225	AA0P WD0FVV/R	11
GA Atlanta	Wednesday	2130 local	145.410	W4BIW W4PME/R	
IL Chicago	Wednesday	1930 local	146.880	WD9IIC K9GFY/R	7
MI Detroit	Wednesday	2000 local	224.460	WD8CIK K8OCL	12
NY Warwick	Tuesday	2000 local	144.280	WA2LQQ	6
TX Houston	Tuesday	2000 local	145.450	WA5ZIB WB5RDK/R	
TX Dallas	Wednesday	2000 local	146.610	WB5PMR ???/R	

Voice Notes

[1] This net may return to 21.280, summer 1987 propagation conditions permitting.
[2] This net may return to 21.280, summer 1987 propagation conditions permitting.
[3] Interim frequency; frequency is ±10 kHz.
[4] WA6YCZ/R; additional links on K6GWE/R, 443.525; W6OA/R, 146.655; KU6A/R, 223.720 MHz.
[5] Two-meter simulcast of South Pacific HF net by W6SP.
[6] Two-meter simulcast of 75-meter East Coast net by WA2LQQ.
[7] PL 1B required for access.
[8] Back-up frequency is 7.064 MHz.
[9] Two-meter simulcast of 20-meter net by ZS6AKV.
[10] From Pietermaritzburg.
[11] Alternate NCS is WD0HHU.
[12] Also linked via 147.22, 443.00, 443.55 and 1288.99 MHz.
[13] Trial basis for spring 1987. See note 2.

Amateur Satellite Communications

Conducted By
Vern "Rip" Riportella, WA2LQQ
P.O. Box 177, Warwick, NY 10990

Working OSCAR—The Basics

The fascinating thing about Amateur Radio satellites is . . . well, quite frankly, there's not just one thing but rather an array of features that make working OSCAR today among Amateur Radio's most fascinating, challenging and rewarding activities. Last time (June 1985 *QST*), we highlighted a few of the many activities and modes to be found on OSCAR. Now let's look a little closer at what it takes to get started using OSCAR satellites.

There are four basic topics you'll need to be familiar with to succeed on OSCAR:

1) When and where to point your antenna (tracking).
2) Basic transponder characteristics (frequencies and modes).
3) Basic station characteristics (transmit power, receive sensitivity).
4) Basic operating practice (band plan, protocol, courtesies).

First, you need to know when and where to look for the satellite (tracking). In essence, you need to know when the satellite you're interested in working is within view of your QTH. If you've got one of the new, little personal computers, AMSAT has a tracking program that makes (tracking) life very straightforward and easy. A review of the ARRL's *Satellite Experimenter's Handbook* (SEH) is recommended for a basic understanding of what's going on with the satellite—that is, its motion around the earth. But, just as you don't need to understand hydraulics to use the automatic transmission in your car, you don't need to understand Kepler's equations to use a computer to track OSCAR. But it helps conceptually to have a basic grasp of the motion of the satellites across the sky.

Several manual means of satellite tracking are also available. These aids are available from ARRL (OSCARLOCATOR), ZRO Technical Devices (Satellipse, Satellabe), and others.[1] The manual devices resemble circular slide rules with maps and various curves on them. When you enter a reference time and position for a certain satellite, the locators will tell you where the same satellite is at the time you select. Significantly, it will also tell you which way to point your antennas (azimuth and elevation) to zoom in on the OSCAR you want to work. Although the manual tracking aids or locators lack the accuracy, speed and general pizzazz of computer tracking methods, they are inexpensive and fairly easy to learn to use.

On the other hand, our first tracking computer was a Sinclair ZX-81, which I purchased for $29.95 plus an extra $10 for more memory. I got the AMS-81 tracking program from the AMSAT Software Exchange for $10 and had an old TV monitor just perfect for the job.[2] So for $50 and no sweat at all, I had a first-rate tracking computer system that gave me high accuracy and did nothing but plot orbits for me.

Second, you need to know what frequency to transmit on and what frequency to listen on. The popular current mode of operation on AMSAT-OSCAR 10 (AO-10) is called

Table 1
OSCAR Operating Modes

Mode	Uplink Band	Downlink Band
A	2 m (145 MHz)	10 m (29 MHz)
B	70 cm (435 MHz)	2 m (145 MHz)
L	24 cm (1269 MHz)	70 cm (436 MHz)
S	70 cm (436 MHz)	13 cm (2401 MHz)
J	2 m (145 MHz)	70 cm (435 MHz)
LU	2 m/23 cm	
K	15 m	10 m

Mode B. That means you transmit your signal to the satellite (uplink) in the vicinity of 435.1 MHz, and you listen for your signal (downlink) from the satellite on about 145.9 MHz. Moreover, since you are operating crossband, you operate full duplex—that is, you transmit and receive concurrently. Why crossband? The hardware required to provide sufficient isolation for in-band, full-duplex operation (similar to a terrestrial FM repeater) would be too massive to fit aboard a cramped-satellite. So the traditional satellite solution is to have cross-band, full-duplex operation. OSCAR satellite modes are listed in Table 1. A Mode B band-plan chart is shown in *The 1985 ARRL Handbook*, page 23-6.[3]

Third, you need to have a basic understanding of how to establish a station to use AO-10 in the mode you choose. You will need information to help you decide what equipment need be added to your existing shack to make it OSCAR-compatible.

Fourth and finally, you need to be familiar with the operating practice and operating schedule for the satellites and modes you choose to operate. Operating schedules can be learned from AMSAT newsletters and nets.[4]

Operating practice can be learned from many reference sources, including the ARRL's *SEH* and AMSAT's beginner's manual.

Each of these topics will be covered in detail in future columns. Meanwhile, here's something you can do to get started right away. Simply take the first small step: Listen for AO-10. Make an attempt to hear AO-10 on its next available pass. All you need is a 2-meter all-mode receiver and antenna. Having a 2-meter preamp is not absolutely essential, but it makes a very, very significant improvement. Borrow or purchase one if you can. Then write or call the AMSAT Area Coordinator in your region and ask for a copy of the latest orbital predictions for your area. (You should offer to contribute to the postage costs.) Ask the Area Coordinator to help you point your beam in the right direction if AO-10 is in view then. Check to see if it's in Mode B or Mode L. You want Mode B if you don't know how your Area Coordinator is, you can find out by calling toll-free to the Chief Area Coordinator, Jack Somers, WA6VGS, at 1-800-421-6631.

If you're lucky, the Area Coordinator will be close by, and a short drive to his QTH will result in your acquiring the key tracking data for your area. The computer-tracking printouts will tell you the azimuth and elevation to point your antenna to track AO-10 for the next several days. Most newcomers don't have an elevation rotor. In that case, you may have some difficulty hearing AO-10 when it's above 20- or 30-degree elevation. As a start, you might select a listening period during which AO-10 is within 20 degrees or so of the horizon.

Start out by listening to the general beacon on 145.810 MHz. The frequency may be slightly different because of the Doppler shift

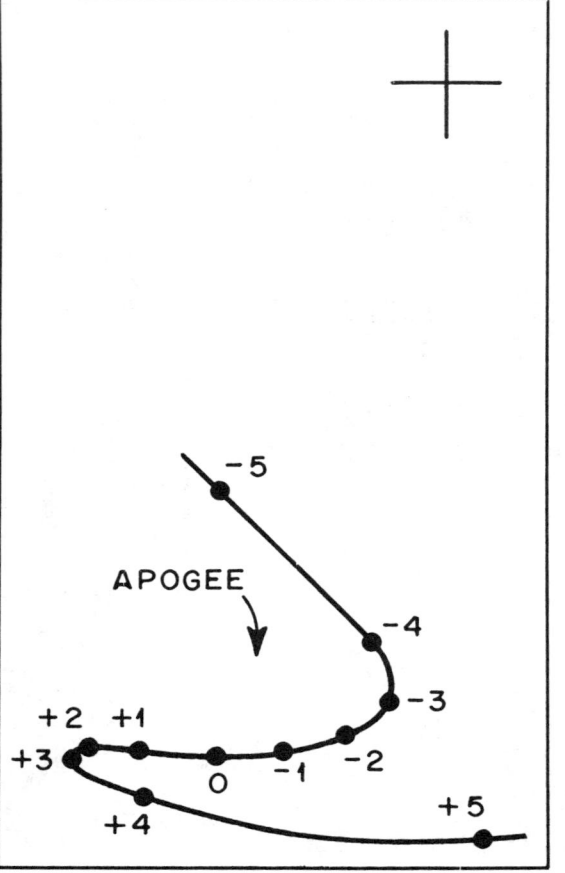

AMSAT-OSCAR 10 ground-track cursor for the OSCARLOCATOR. Reference data is for September 1.

effect exhibited by the fast-moving satellite. Use your receiver's USB or CW mode to listen. You may hear a buzzing sound, which is special phase-shift-keyed (PSK) telemetry, or you may catch the CW or RTTY bulletins, depending on exactly when you listen. The beacon changes mode in recurring half-hour patterns (beacons will be detailed in a subsequent column).[5]

Chances are, if you don't use a preamplifier, you will hear the beacon weakly or not at all. A good preamp is really very helpful, especially if your antenna is not terribly good or is in poor repair. Also, if you have a long run of transmission line between your antenna and your rig, the preamp will help most if mounted at the antenna and remotely powered from the shack. For a short test, you might consider battery powering the preamp and locating it at the antenna, just to see how well it works that way.

In any case, try tuning up to center band, 145.900 MHz. If the transponder is on, you should hear dozens of CW and SSB signals in the transponder passband. Use the USB position to receive. The convention on all the transponders is to use an uplink mode that produces USB on the downlink. Since AO-10 Mode B is an inverting linear transponder, LSB on the uplink results in USB on the downlink.

If your 2-meter station consists of a 2-meter beam with at least 10-dB gain and you have a preamp, if the transponder is on, you'll be hearing lots of activity from local as well as DX stations. And at this point, you will have linked yourself for the first time into the worldwide network of satellite users.

But if you're disappointed in what appear to be weak signals on the 2-meter downlink, take heart. There really *is* a lot of 2-meter signal there, but your everyday, off-the-shelf 2-meter rig may need some help with the AO-10 signal. Most of the 2-meter radios sold have only marginal receiver front ends and really perk up with the addition of a new GaAsFET preamp out front. "Wow," you'll exclaim, "What a difference!"[6]

It is often helpful to a have a standard frame of reference or benchmark to determine just how well you *are* hearing on 2 meters. To provide just such a benchmark, AMSAT began a series of on-the-satellite tests last spring. The tests are similar to the ARRL's frequency-measurement tests (FMT), except AMSAT's test works in the amplitude domain while ARRL's obviously works the frequency domain. AMSAT's receive sensitivity tests offer a convenient way to establish how well your station is performing both in an absolute sense and compared to other operators. For those who demonstrate various levels of proficiency on the test, an AMSAT Technical Achievement Award, called the ZRO Memorial Station Engineering Award, is available. The test frequency is 145.840, and the next scheduled event will be later this month.[7]

First steps in a new endeavor are often long remembered. Many new OSCAR enthusiasts assert that their first OSCAR QSO was even more exciting than their first-ever QSO. Of course, you'll have to find that out for yourself. Meanwhile, your portable ionosphere whirs overhead, awaiting your signal. Once you hear OSCAR's beckoning call, you'll be hard pressed to resist for long!

Next time, we'll take a look at the antennas used in OSCAR work.

Notes

[1]ZRO Technical Devices, P.O. Box 11, Endinott, NY 13760.

[2]AMSAT Software Exchange, P.O. Box 27, Washington, DC 20044.

[3]A copy of the Mode B band plan is available free for a business-sized s.a.s.e. to the author at the address at the top of this column.

[4]See notes to the June 1985 column for nets and newsletter information.

[5]A free handout showing the general beacon program schedule is available from this column conductor for a business-sized s.a.s.e.

[6]Ads for 2-meter preamps are found in this *QST*. Refer to the ad section.

[7]A free information booklet about receive sensitivity and the AMSAT ZRO-Memorial Technical Achievement award is available for a business-sized s.a.s.e. with 39¢ postage affixed. Request from this column conductor at address above.

Amateur Satellite Communications

The Digital Satellite World

Next year, AMSAT's Phase 3C spacecraft will carry four transponders, one of which is the so-called RUDAK digital transponder. High in its elliptical orbit, Phase 3C and RUDAK may serve as a digital trunk for terrestrial networks. The following description of RUDAK comes from our German colleagues at AMSAT-DL. In particular, a team in Munich under Hanspeter Kuhlen, DK1YQ, has built RUDAK. Read now what it's all about.

RUDAK Status Report of the RUDAK Group of AMSAT-DL

By Peter Guelzow, DB2OS
Deputy RUDAK Project Leader
(Translated by Don Moe, KE6MN/DJ0HC)

"RUDAK" stands for "Regenerative Umsetzer fuer Digitale Amateur Kommunikation" (in English: Regenerating Transponder for Digital Amateur Communications). It is comparable to a so-called digipeater (digital repeater). Digipeaters are terrestrial relay stations for packet radio. They relay digital information between two stations in case there is no direct path between them.

Similarly to analog transponders, it seems desirable to install such a digipeater at the highest possible location with a large coverage area, eg, aboard a satellite in earth orbit. Thanks to the highly elliptical orbit of Phase 3C, RUDAK should eventually enable the interconnection of several local area nets in addition to point-to-point contacts between radio amateurs across the entire world. Naturally, a relay station with such a large coverage area has to contend with a series of difficulties. For example, the problem of multiple uncoordinated access or the selection of optimal modulation techniques are only two of among many that could be mentioned. These and other problems are to be researched primarily with the help of RUDAK with the goal of developing suitable techniques and protocols which will benefit future projects.

The initial designs of the RUDAK experiment were determined at a working meeting at AMSAT-DL in Marburg, West Germany in February 1985. In July 1985, in Marburg, the entire hardware design, the IHU interface as well as the satellite interface were presented. After certain modifications were agreed, the first functional wire-wrap prototype, RUDAK 1, was unveiled September 6-7. At this meeting in Marburg the primary task was to integrate the programming language IPS, previously developed by Dr Karl Meinzer, DJ4ZC, into the RUDAK processor. After several software errors were eliminated, IPS-CR was at last successfully loaded into RUDAK. The successful implementation of the IPS system brought the RUDAK experiment a giant step closer to completion.

The first printed circuit version, RUDAK 2, supplemented the original wire-wrap version a short time later. In all, the plan calls

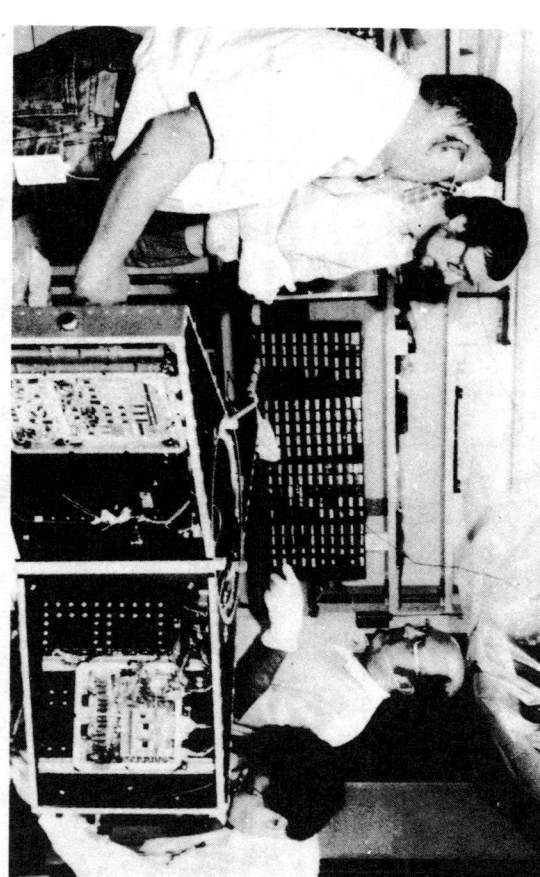

Part of the AMSAT-DL team during the Phase 3C integration in Golden, Colorado recently. Shown (l-r) are Konnie Mueller, RUDAK Project Manager DK1YQ, DJ4ZC and DJ5KQ.

for four double-sided circuit boards with plated-through holes with the dimensions of 290 × 180 mm. Two boards will be built as identical flight versions, with one serving as a reference model on the ground. The other flight version will be mounted together with the demodulator and the power supply in a two-section housing with the dimensions of 300 × 200 × 20 mm and 300 × 200 × 17 mm. This will be subsequently integrated into the Phase 3C satellite. The remaining circuit boards are reserved for software development and various tests such as radiation testing. The boards were laid out using a CAD/CAM system.

The hardware development of the RUDAK processor is completed. The main work now involves the completion of the flight version as well as the implementation of the AX.25 protocol.

On January 24-25, 1986, the RUDAK group met once again in Marburg to clarify remaining details regarding integration into the satellite. A further high point was the demonstration of RUDAK's capabilities. For the first time, four TNC1s were linked together via the RUDAK processor, simulating on hard-wire connections how the operation will later take place. TAPR TNC1s were used exclusively, though only one had the original TAPR software; the other three used the multi-connect firmware from WA8DED. A lively data exchange took place, and DJ4ZC made his first packet-radio QSO. Additionally, RUDAK transmitted some general information in beacons. As was to be expected, numerous collisions occurred. Even so, RUDAK demonstrated that it already was working correctly. The next milestone is May 10, when the RUDAK flight version has to be ready for integration into the Phase 3C satellite.

The RUDAK hardware consists of 25 integrated circuits and only two discrete transistors. The entire circuitry was realized using CMOS technology, so power consumption is only 300 milliwatts. The heart of the RUDAK processor is the CMOS version of the 6502 CPU, which is clocked at 800 kHz. For storage of the RAM-resident system software and data, 56 kbyte of static CMOS RAM chips are provided. This concept itself gives RUDAK greater flexibility in case, for example, the entire RUDAK software has to be updated due to changes in the protocol, as has already been practiced with OSCAR-10's IHU. A single 2-kbyte fusible link CMOS PROM is used to load the IPS system via the boot PROM contains various programs that will perform tests of the entire hardware in the RUDAK processor while in orbit.

To communicate with the outside world, the RUDAK processor has various parallel and serial input/output ports. One serial line and one 8-bit parallel port with the appropriate control lines are used for communication with the IHU. In the start-up phase, these paths are used to transfer diverse command and diagnosis instructions. Later, using this same path, RUDAK can receive current telemetry data which can be processed further. The IHU can also use a portion of the RUDAK memory as virtual memory in which to store larger quantities of data, eg, RTTY/PSK bulletins. The capacity of the 16 kbyte of RAM in the IHU is already totally used.

Normal operation with ground stations is handled by the RUDAK packet port. One send and one receive channel are available. The heart of this port is the CMOS version of the Z80-SIO, a universal chip that supports the AX.25 protocol in addition to asyn-

Conducted By
Vern "Rip" Riportella, WA2LQQ
PO Box 177, Warwick, NY 10990

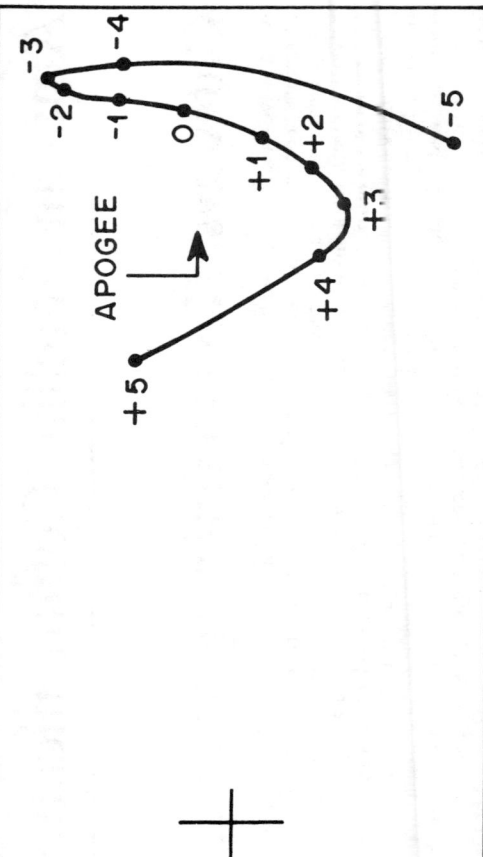

AMSAT-OSCAR 10 ground-track cursor for the OSCARLOCATOR. Reference data are for August 15.

chronous and synchronous operation.

An independent receiver in the Mode-L transponder is provided for the RUDAK uplink on 1269.675 MHz. The demodulator converts the 2400-bit/s biphase PSK signal into a clean digital signal for the RUDAK processor. Thanks to the sweep circuit in the demodulator, the uplink signals only have to be in the capture window within plus/minus 7.5 kHz of the center frequency.

On the downlink side, the output data modulates the RUDAK beacon transmitter in the L-transponder on 435.675 MHz using BPSK at a data rate of 400 bit/s; the same as for the general beacon of OSCAR 10. Experimentally, the rate can be increased to 1200 bit/s using NRZI modulation.

The 2400-bit/s uplink and the 400-bit/s downlink signals are generated using the AMSAT standard, just the same as for the general beacon of AO-10. In the AMSAT standard, the data bits are transmitted differentially, ie, a logical "0" is sent when there is no change in two successive bits, whereas a logical "1" is sent for a change between bits. Additionally, the clock signal is combined with this data stream. Due to this trick and the differential encoding, the design of the decoder is significantly simplified.

Unfortunately, another standard has established itself internationally in which the assignment of the logical levels is exactly reversed. In the NRZI standard, a logical "1" is transmitted when there is no change between bits. If the bit clock is also combined with the data, the signal is then called "NRZIC." In order to reduce the confusion as much as possible, it was decided to adopt the previous AMSAT standard for RUDAK. In the case of the 1200-bit/s downlink option, the NRZI standard was chosen, and, in contrast to the AMSAT technique, the clock signal is not combined with the data, since to do so would exceed the bandwidth of the SSB receiver.

In the initial stages, RUDAK will emulate the existing digipeater functions as they are defined the AX.25 protocol version 2. No mailbox operation is planned presently, although various other messages, such as bulletins, orbital data, telemetry values and user instructions, can be cyclically transmit-

ted when no uplink signals are being digipeated. New ground stations can take their time in adjusting their receiving equipment.

Additionally, a robot-type operation is planned in which the ground stations "connect" to the satellite and are assigned a consecutive number. In a fashion similar to the RS satellites, a RUDAK command station could later download the list and send out QSL cards. It is also hoped that an overview of packet-radio activity worldwide could be thereby obtained. Should a suitable link-layer level 3 protocol subsequently become available, it could possibly be implemented.

For the majority of the terminal node controllers, eg, TAPR TNC1, AEA PKT-1 or Heath HD-4040, the only software modification required is an updated EPROM to handle a hardware bug in the WD1933/35 HDLC controller. Otherwise, only a PSK modem for 400/2400 bit/s has to be connected to the external modem jack in the TNC. Other TNCs, such as the Kantronics "Packet Communicator" or various software solutions, are unfortunately not suitable due to the software and/or hardware restrictions.

The TNC must be capable of operating full-duplex at different transmit/receive baud rates and support the connection of an external modem.

Besides the normal equipment, a so-called "RUDAK User Interface" is required. This is under development by the RUDAK group and AMSAT-DL. The RUDAK User Interface consists of a converter that translates a 2-m signal to 24 cm and modulates the carrier with 2400-bit/s BPSK, and the "AMSAT-AFREG," which is the BPSK demodulator for the 400-bit/s downlink. Additionally, various buffers and controls for switching the different signal paths and a power supply are needed. The various schematics, especially for the AMSAT-AFREG and the converter, will be published by AMSAT-DL after the design is completed.

On the RF side of the ground stations, the 400-bit/s downlink signal on 435.675 MHz should provide a signal strength of 12-dB Eb/No to an antenna with 10-dBi gain. For the uplink on 1296.675 MHz, 12 watts (11 dBW) into a 15-dBi antenna should be sufficient.

Amateur Satellite Communications

Gateways: Keys to Opening New Communication Doors

A gateway is a portal between two domains or regions. For example, in computer networks, a gateway is a facility where different networks meet. Data from one can be transferred to the other. In Amateur Radio, packet-radio gateways illustrate the concept. A station having transceive capabilities on both VHF and HF, and the equipment to switch traffic, could function as a gateway. Thus, the packet gateway functions as a portal to VHF networks for HF network users, and vice versa, as shown in Figure 1.

Similarly, in Amateur Space Program jargon, a gateway is a facility where a terrestrial network interfaces a space network.[1] For example, a terrestrial repeater with its user community could interface AMSAT-OSCAR 10 and its network of satellite users, the Amateur Radio space-communications community. The facility providing this network interface is a gateway. A typical satellite gateway facility might look as simple as Figure 2.

Conceptually and functionally a satellite gateway is straightforward. Typically, signals originated by 2-meter FM users are converted to baseband audio by the repeater's receiver. Then, by any one of several means, the audio is linked to the satellite station's uplink transmitter. Conversely, satellite downlink signals are received at the ground station, converted to baseband audio and then shipped back to the repeater's transmitter. From there they go out on FM just as if they were a normal repeater signal.

The overall system (including the gateway, the repeater, the satellite ground station and the users in both the terrestrial and space communications networks) is functionally identical to normal satellite systems used by individuals, except the user on one end has an extra link inserted to "remote" him from the satellite station. Indeed, regular satellite users might be unaware they were talking to someone quite apart from a regular satellite station. You might be strolling down the street in sunny Orlando, Florida chatting with your 2-meter hand-held radio and chatting with a chap in Honolulu or Sardinia. And, except for the repeater squelch tail, no one would suspect the unique nature of the QSO in progress. But dozens of these types of gateways have operated since AO-10 was launched in 1983. Two years ago, the first transcontinental QSO via AO-10 using hand-held radios took place when two gateway stations hooked up and hand-held users in West Virginia and California QSOed through the gateways in their vicinity.

Two types of gateways have been used with AO-10. The first and most basic type provides single channel access through a local repeater on one end and a fairly standard AO-10 ground station on the other end. The second type of gateway is slightly more complex. It takes a few dozen kilohertz at one frequency and translates it to the AO-10 uplink frequency. The uplink could contain several mixed SSB and CW signals spread across, typically, 40 kHz. This rarer type of gateway uses what is called a linear translator.[2] Yet again, AO-10 users might be totally unaware that the person they are talking with is using a gateway to access AO-10.

Gateway operation through AO-10 requires skilled operators; completely automatic control is still in the future. With AO-10, antennas must occasionally be aimed and Doppler shift corrections must be made. While the control operator of a simple gateway has only to manage a 2.5-kHz chunk of AO-10 spectrum, linear translator operators need be much more skilled. The relative challenge is comparable to the difference between steering a canoe and a barge through a narrow channel.

The advantages of using a gateway to access AO-10 are many. Apartment dwellers unable to field the modest antennas required for direct AO-10 access can still enjoy the occasional thrill of working true DX by using a gateway. Mobile operators can enjoy intercontinental QSOs on VHF even before the new generation of satellites is born.[3] Demonstrations at conventions and hamfests might be greatly enlivened with the addition of a gateway operation to show off Amateur Radio's "high ground." Emergency communications might be enhanced if a trunk line to the affected area passed via a gateway and AO-10 to master command centers and logistics-support areas. Many more classes of uses might be imagined.

But one of the most important aspects of gateway operation is the facilitation of demonstrations to the uninitiated of the thrill and challenges presented by today's Amateur Radio in a space context. Moreover, if you surveyed hams today, you'd probably find that if they had just one radio, it would likely be a 2-meter hand-held radio. Thus, with a gateway to use, a newcomer with hamdom's most basic equipment, a hand-held radio, could try out hamdom's highest achievement—AO-10! *That* is a significant capability bridge that should further spur gateway use.

Yet some veteran satellite users have expressed dismay and concern with some aspects of gateway operation. The concerns most often fall in one of two categories. The first is a turf issue. "They'll use up all the spectrum and available power." The second is more subtle. "If gateways are really good,

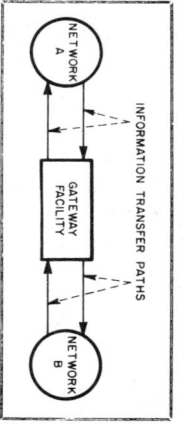

Fig 1—The gateway as a bridge between networks.

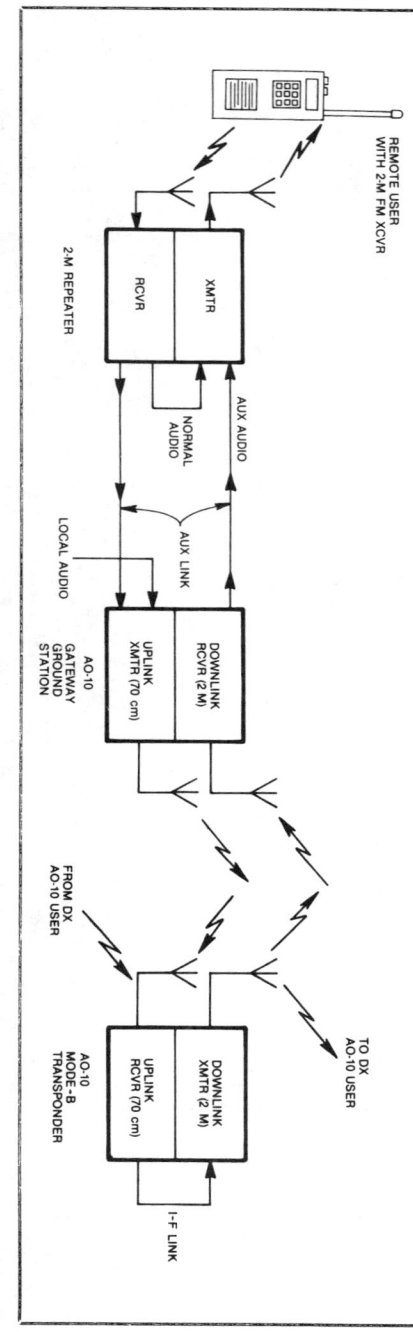

Fig 2—Gateway functional block diagram. The aux link between the repeater may be any convenient mode such as telephone line or aux RF link. Even 24-GHz Gunnplexer transceivers have been used.

there'll be no reason to obtain one's own satellite station, people will ignore the desired space education aspects of getting to know how to track and use satellites, and so forth. And if they don't need to know anything, they won't need AMSAT to provide information. AMSAT membership will drop and it won't be able to build satellites anymore!"

Fortunately, none of this has happened. In the first case, the number of gateway operations has not been excessive. The very fact that AO-10 does not always present itself for use at convenient times has tended to limit gateway operation since few find it sufficiently compelling to warrant rising at, say, 3 AM.

On the second issue—the matter of diminished AMSAT membership with increased gateway use—again, fortunately the issue has remained mostly academic because gateway operations have not proliferated. Moreover, many gateway users have found AO-10's challenge so appealing they have obtained both their own OSCAR equipment and AMSAT membership to boot.

In addition, the built-in constraints on gateway operation seem to work to encourage truly interested individuals to do more, to learn more. Individuals accessing the satellite through a gateway often find it so fascinating they want to get more flexibility in who they talk with on the satellite and when they operate through it. This strongly impels them toward obtaining their own stations and to become full-fledged AMSAT members. Those who remain occasional gateway users are perhaps content to queue up for a short QSO on AO-10 through the gateway and that's that. These folks have had fun, enjoyed the tryout and will probably tell friends about it. They can continue to be AMSAT's guest on the satellite even though we hope they will eventually help to support new satellite construction by joining.

Gateway operation is a great way to taste the wine before purchasing the bottle. And it offers some nontrivial benefits and experience to many who try it. See if there is an experienced AO-10 operator in your area and a cooperative, knowledgeable repeater operator, too. You might suggest that these folks get together for a gateway experiment. Your friends may never forget the experience of their first AO-10 contact made from the comfort of their "whatever"![4]

Next month, we'll take a first look at Japan's first homebuilt Amateur Radio satellite, JAS-1. Employing both digital and voice transponders, JAS-1 will be launched late this summer.[5]

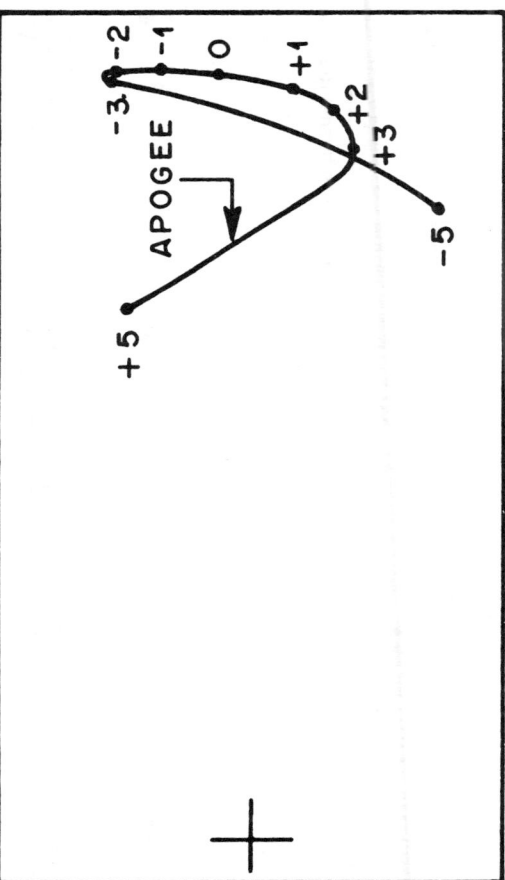

AMSAT-OSCAR 10 ground-track cursor for the OSCARLOCATOR. Reference data is for May 15.

Notes

[1]The term "teleport" is also applied to the interface of space and terrestrial networks.

[2]Functionally, a linear translator is very similar to the linear transponder employed by AO-10.

[3]See last month's column on prospects for mobile satellite work.

[4]A revised booklet on gateway operation is available for a $5 donation to AMSAT. Checks only, please, made payable to "AMSAT." Send to AMSAT, PO Box 27, Washington, DC 20044. Mark your envelope "Gateway" to speed your request.

[5]Free information on satellite-tracking software, AMSAT satellite nets and how to get started on the satellites can be obtained by sending a business-sized SASE (with postage commensurate with your request) to the column conductor at the address above.

Amateur Satellite Communications

Conducted By
Vern "Rip" Riportella, WA2LQQ
PO Box 177, Warwick, NY 10990

Introducing Phase 3C: Newest High Flyer Debuts Soon

If all goes well in the next few weeks, Amateur Radio satellite enthusiasts the world over will soon be enjoying the very latest in OSCARs. After a complicated series of launch delays spanning more than a year, AMSAT's newest and most powerful OSCAR ever sits poised and ready for action. What is it all about? How can the newest OSCAR be used? What can you expect from this satellite? These are some of the questions I'll address in a series of columns on the Phase 3C bird beginning this month. Subsequent columns will fill in the details. This first installment will give you a quick overview.

First, the Basics

Phase 3C is an OSCAR, an Orbiting Satellite Carrying Amateur Radio. The Phase 3 series, first conceived in the mid-'70s, comprises long-lived spacecraft placed in high, elliptical orbits. The C version is the third in the series.[1]

The elliptical orbit sought is a slight modification of the Molniya-type orbit used by Soviet communication satellites for years. The Soviets use the Molniya orbit because much of their territory is at high latitudes. Geosynchronous satellites, which by definition must be situated 22,300 miles above the equator, serve extremely high latitudes poorly. This is because from such latitudes, the geosynchronous satellites appear low on the horizon. That poses siting problems, and requires large-aperture antennas to focus on the satellite while rejecting noise and multipath effects from the ground.

Radio amateurs have used Molniya-type orbits for other reasons, however. The Molniya orbit can be thought of as semi-synchronous. For much of its orbit, the satellite is very high, about the same as a geosynchronous satellite. Moreover, its movement across the field of view of well-positioned ground stations is small for much of the orbit, especially around apogee, the high point of the orbit. But as the satellite approaches perigee, the orbit's low point, it quickly sweeps around the earth and out of view.

A series of Molniya-orbiting satellites could serve the Amateur Radio community much as a series of geosynchronous satellites. But, given the economic realities of building satellites, a fleet of coordinated Phase 3 birds is unlikely. And a single Molniya-orbiting OSCAR is far better than a single geosynchronous OSCAR as far as sharing the resource with all amateurs. Depending on the exact orbit attained, Phase 3C should provide hours of great DX daily for hundreds of users.

What Kind of DX?

In contrast to AO-10, which has a

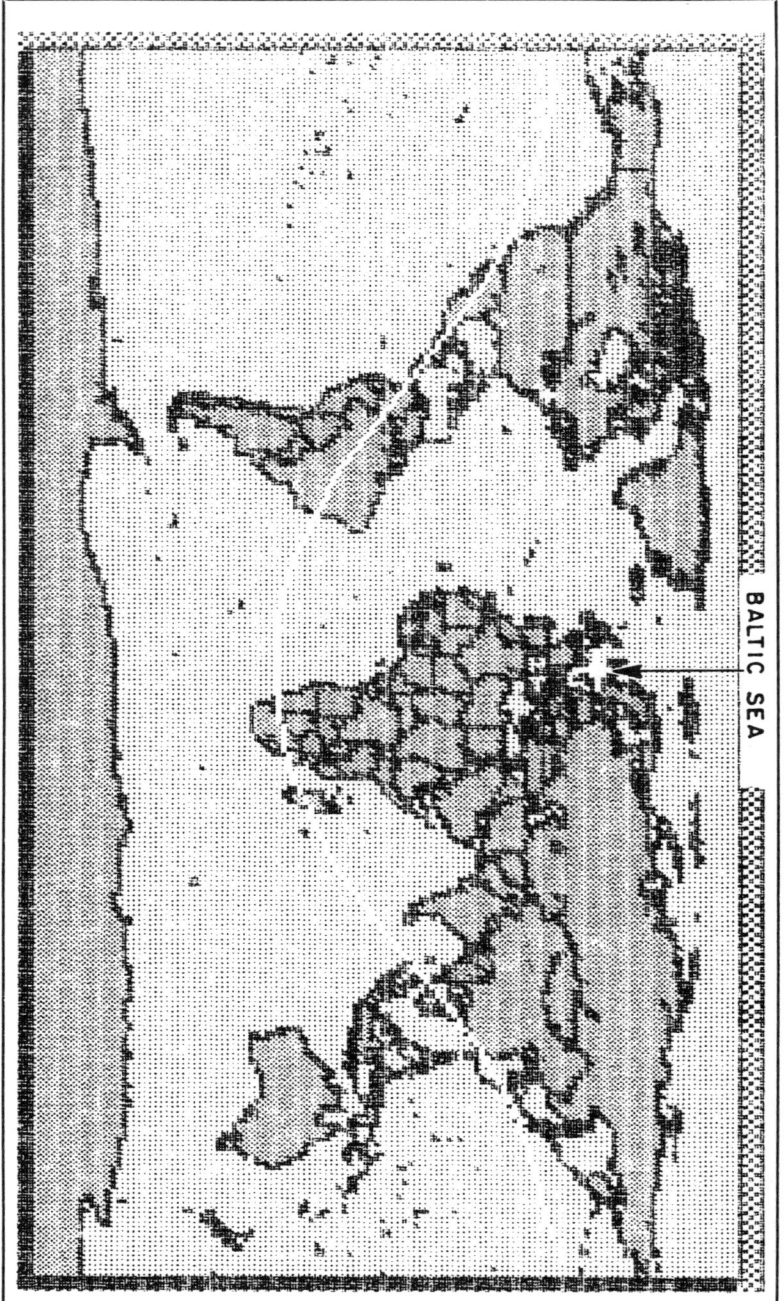

Fig 1—Typical Phase 3C "footprint." The footprint of a satellite includes all locations that can see the satellite, ie, are line-of-sight with it. All stations within the footprint can communicate with each other through the satellite. Here, the footprint is the area above the broad, apparently irregular, curve covering (mostly) the northern hemisphere. In fact, the footprint appears irregular only because of the distortions inherent in a Mercator-projection map.
In this rendering, the satellite is positioned over the Baltic Sea where the white cross is the sub-satellite point, or point directly beneath the satellite. The serpentine curve is the border of the footprint. From the satellite, the curve marks the earth's visible limb.
The footprint diameter varies with satellite height. Phase 3C's height will vary over approximately a 25:1 range in its planned ellipse. On occasion, Phase 3C's footprint will blanket as much as one-quarter of the earth's surface.
This is a section of a printout made using Graftrak II, a commercially available product of Silicon Solutions, which runs on IBM® PC or compatible computers only. For more information, contact Silicon Solutions, PO Box 742546, Houston, TX 77274-2546, tel 713-777-3057.

26-degree orbital inclination, Phase 3C satellites are designed to operate in a 58- to 60-degree inclined orbit. Thus, when apogee occurs over northerly latitudes, a station in the US can expect to work a huge stretch of territory as far east as Indochina (see Fig 1). Apogee will be about 22,000 miles, perigee about 700 miles. DX vistas will vary widely. However, from virtually any spot on earth, communications windows to rare DX locations will appear in long and short cycles.

Phase 3C is a communications satellite. It carries transponders, power-generating and conditioning equipment, attitude-adjustment equipment, a propulsion system and a command/telemetry computer.

Transponders are similar to terrestrial repeaters, but with some important differences. For example, all four transponders on Phase 3C operate cross-band (see Table 1). Cross-band operation is necessary to provide the desired suppression of the transmitted signal in the receive passband when operating a collocated receiver.[2]

Another significant difference between repeaters as used in terrestrial operation and satellite transponders is their bandwidth. Phase 3C will provide a passband several hundred kilohertz wide. Typical repeaters have a passband usually less than 10 kHz wide. Moreover, the Phase 3C transponders are linear repeaters. That means any modulated carrier heard on the input receiver (uplink) will be reproduced faithfully by the transmitter (downlink). Essentially, what goes in, comes out. Terrestrial repeaters usually run class C for efficiency reasons. FM passes through such repeaters very well; SSB not well at all.

For reasons of power conservation, the allowed modes of operation on OSCARs include SSB and CW, but not FM. Various special modes are also permitted. For example, although FSK packet radio is marginally acceptable, a very strong emphasis is being placed on using more efficient packet-radio modulation techniques. Primary among the preferred modulation techniques is phase-shift keying (PSK). PSK is used by AMSAT OSCAR 10's telemetry system, and is used on Fuji OSCAR 12's Mode JD (digital) downlink. Phase 3C's telemetry system will use PSK as well. PSK offers a huge signal-strength (at least 10 dB) advantage over FSK. SSTV and AFSK RTTY are marginally acceptable for Phase 3C operation. The reason these modes are only marginally acceptable is that they waste so much of the satellite's power compared to the information content. In particular, AFSK RTTY has been thoroughly eclipsed by packet radio as a digital communications mode. For similar power conservation reasons, FM voice operation is *prohibited entirely* on OSCARs. FM (and full-carrier AM for that matter) expend power even without modulation. Thus, pauses between words during speech *waste power*. In sum, SSB, CW and PSK are the recommended modes of operation on Phase 3C.

A final difference between Phase 3C's transponders and terrestrial repeaters is that the passbands are inverted. That is, a rise in frequency in the uplink results in a fall in frequency in the downlink. Similarly, LSB transmitted on the uplink results in USB on the downlink.[3] Aside from some arcane mixing reasons for designing inverting transponders, this scheme reduces Doppler shift to a small degree.[4]

Next month I'll detail the station equipment you'll need to work any of Phase 3C's four transponders.[5]

Table 1
Phase 3C Modes and Bands

Mode B:	70 cm up; 2 meters down
Mode JL:	24 cm and 2 meters up; 70 cm down
Mode S:	70 cm up; 13 cm down
RUDAK:	24 cm up; 70 cm down
Beacons: (MHz)	Mode B: general beacon 145.8125; engineering beacon 145.975
	Mode JL: general beacon 435.650; engineering beacon 435.675
	Mode S: 2400.640

Notes

[1]Phase 3A was lost because of a launch-rocket failure on May 23, 1980. Phase 3B was launched successfully on June 16, 1983, and is now known as AMSAT OSCAR 10. Tradition dictates that a satellite takes its name only after becoming operational.

[2]Repeaters typically use duplexers to achieve the required isolation of transmitted and received signals separated only by a percent or two of the operating frequency. The extremely high Q of cascaded cavities is necessary to notch out the transmitted signal from the receiver input. Cavities are bulky and heavy, and are impractical to carry on satellites. This dictates the need for cross-band operation.

[3]Operating convention prescribes that all downlinks should be USB. Thus, uplink signals to inverting transponders must be LSB; uplink signals to noninverting transponders must be USB.

[4]Doppler shift can vary from a few hertz to a few hundred hertz per minute on Phase 3 satellites.

[5]Send an SASE to AMSAT, PO Box 27, Washington, DC 20044 for information on how to get started on Phase 3C and on other satellites, and how to become an AMSAT member.

Amateur Satellite Communications

Conducted By
Vern "Rip" Riportella, WA2LQQ
PO Box 177, Warwick, NY 10990

Introducing Phase 3C: "Superbird" Soon!

Soon, the most complex piece of Amateur Radio hardware ever built will be lifted into a high, elliptical orbit. This hardware will provide enormous communications capabilities for those equipped to take advantage of it. But what does it take to get "on board" the new Phase 3C satellite? This month we'll look at a typical "starter" station for Phase 3C.

As detailed last month, Phase 3C[1] will employ four distinct modes of operation: B, JL, S and RUDAK. Each mode refers to a set of uplink/downlink band pairs and/or modulation types (see Table 1).[2]

To operate Mode B, you need to know five key facts:

1) Where the satellite is
2) When its Mode B transponder is on
3) What frequency to transmit on
4) What frequency to receive on
5) Basic operating practice

Determining the satellite's location, item 1, is called *tracking*. Since high-gain antennas are required, you need to properly aim them at the satellite. Tracking tells you *where* to aim. Tracking techniques and equipment have been covered in this column previously, and will be repeated in an upcoming column.

Item 2 is a matter of scheduling. The operating schedule of Phase 3C will be announced after launch. You can learn the operating schedule by tuning to W1AW bulletins, or by obtaining official AMSAT net bulletins via on-air nets,[3] packet-radio bulletin boards or from telephone bulletin boards.

Items 3 and 4 are addressed in Table 2. Although the table information might lead you to conclude that the passband is divided into many 10-kHz channels, this is not the case. For example, Table 2 shows that if your uplink is on 435.505 MHz, your downlink will be about mid-band on the 2-meter downlink, 145.895 MHz. But, you could just as well have transmitted on 435.507 and heard your downlink on 145.893 MHz. Table 2 merely shows the correlation of uplink and downlink frequencies across a continuous spectrum of available frequencies. Thus, there are no "channels" as such.

There is more to operating via an OSCAR than just turning on the transmitter, though! Item 5, operating practice, will be covered in a future installment. However, well before you turn on the rig, you need to know your basic station requirements to have a good chance of success on OSCAR. So, let's spend a few moments reviewing basic Mode B station requirements. Then you can determine how your shack stacks up and determine where any improvements need to be made.

Let's start at the interface to the satellite: the antennas. You'll need an uplink antenna for 70 cm (435 MHz) and a downlink antenna for 2 meters (145 MHz). How much gain should the antennas have? For the uplink, you want to aim for an effective isotropic radiated power (EIRP) of 27 to 30 dBW (0.5 to 1.0 kW EIRP). EIRP is calculated simply by adding the gain of your antenna in dBi to the power you feed to the antenna in dBW (dB relative to 1 W). Table 3 shows a conversion of watts to dBW. Depending on how much power you have in the shack to feed to the antenna, you can now narrow the range of antennas from which to select. As a rule of thumb, select a 70-cm antenna with at least 10 dBi gain. About 17 dBi gain is the upper limit of gain for practical Yagi or LP antennas used for OSCAR satellites.

Because the satellite receives and transmits using Right Hand Circular Polarization (RHCP), you'll get best results when your antenna is similarly polarized. A

Table 1
OSCAR Operating Modes

Mode	Uplink Band	Downlink Band	Notes
A	2 m (145 MHz)	10 m (29 MHz)	Traditional
B	70 cm (435 MHz)	2 m (145 MHz)	Current favorite
L	24 cm (1269 MHz)	70 cm (436 MHz)	First use: AO-10
S	70 cm (436 MHz)	13 cm (2401 MHz)	New; Phase 3C use
JL	2 m & 24 cm	70 cm (435 MHz)	New; Phase 3C use
JA	2 m (145 MHz)	70 cm (435 MHz)	FO-12 analog mode
JD	2 m (145 MHz)	70 cm (435 MHz)	FO-12 digital mode
K	15 m	10 m (29 MHz)	RS-10/11 use
T	15 m	2 m (145 MHz)	RS-10/11 use
KT	15 m	10 m & 2 m	RS-10/11 use
KA	15 m & 2 m	10 m (29 MHz)	RS-10/11 use

Table 2
Mode B Uplink Frequency Versus Downlink Frequency
(Preliminary Estimates)
(Frequencies in MHz)

Uplink	Downlink	
	145.975	—Engineering Beacon
435.425	145.975	Passband limit, upper
435.435	145.965	
435.445	145.955	
435.455	145.945	
435.465	145.935	
435.475	145.925	
435.485	145.915	
435.495	145.905	
435.505	145.895	—Passband center
435.515	145.885	
435.525	145.875	
435.535	145.865	
435.545	145.855	
435.555	145.845	
435.565	145.835	
435.575	145.825	—Passband limit, lower
	145.8125	—General Beacon

Table 3
Converting Watts to dBW

dBW = 10 log (P), where P is expressed in watts

P (Power in watts)	Equivalent in dBW (Decibels relative to 1 W)
1	0
2	3
5	7
10	10
20	13
50	17
100	20
200	23
500	27
1000	30

Example: Assume you're feeding 50 W to a 12-dBi-gain antenna. What is the EIRP? Solution: From Table 3, 50 W equates to 17 dBW. Add this to the gain of the antenna in dBi to get: 17 dBW + 12 dBi = 29 dBW EIRP = 794 W EIRP. Similarly, 100 W to a 13-dBi-gain antenna, fed through a 3-dB-loss transmission line yields: 20 dBW − 3 dB + 13 dBi = 30 dBW = 1000 W EIRP.

crossed-Yagi array with between 18 and 20 elements in each plane is the most common solution to the uplink antenna question. A linearly polarized Yagi can also be used, but the compromise will be evident in reduced performance in terms of increased fading, which tends to reduce readability.

For the Mode B 2-meter downlink, your antenna should have between 11 and 15 dBi gain with RHCP. Sure, you can get along with linearly polarized antennas, but fading could become bothersome. Most successful satellite operators use 2-meter crossed Yagis with between 7 and 11 elements in each plane for a total of between 14 and 22 elements. The longer-boom 2-meter antennas are generally very good performers, but elevating an antenna with a boom length of up to 20 feet can be difficult in some installations.

After many years of experimenting, I'm convinced it's a mistake to try to stack circularly polarized, crossed-Yagi antennas. Stacking helices is hard enough. Stacking linearly polarized antennas is a well-developed technology. Stacking crossed Yagis is so complex and sensitive to changes in frequency, polarization, spacing and power division that I strongly recommend you don't try it.[4] The best way to obtain more circularly polarized gain is to increase the antenna's boom length.

Before we move into the shack, there are a few more things we need to mention in conjunction with the antennas. An az-el (azimuth-elevation) rotator system is highly desirable, but not absolutely essential. If, for example, you're using fairly low-gain antennas with their correspondingly broader beamwidths, you can often get along without an elevation rotator as long as the antenna is set at a fixed elevation of some predetermined value, say 30 degrees above the horizon. That way, if your antenna has a half-power beamwidth of 60 degrees, which corresponds to a gain of about 10.5 dBi,[5] the signal at the horizon will be only 3 dB down from the main lobe. Similarly, the signal from the satellite at a 60-degree elevation will be down only 3 dB from the main lobe. Of course, you will need an azimuth rotator nevertheless.

Using a mast-mounted 2-meter preamplifier is very helpful for 2-meter Mode B downlink reception. If you use low-loss transmission line, you can probably put the preamp in the shack, out of the weather. In any case, you *should* use a preamp.[6]

In the shack, a Mode B station needn't be terribly elaborate. Essentially you need a good, stable SSB/CW-capable 2-meter receiver, and a clean 435-MHz SSB/CW transmitter and amplifier. I still use the setup I had 10 years ago for AMSAT OSCAR 7 Mode B: An HF transceiver (TS-820), a 10-meter-to-70-cm transverter (Microwave Modules MMT432-28S) and a 2-meter all-mode transceiver (TS-700S) form the central elements. The transverter feeds a solid-state 70-cm amplifier. If you've accumulated the usual assortment of radios typical of many modern shacks, perhaps you already have most of the gear you need. Otherwise, you may want to investigate several approaches to configuring a successful Mode B station.

Next month, I'll discuss Phase 3C operating schedules. In the meantime, you might want to tune in on the Phase 3C launch. It will be transmitted on a world-wide AMSAT Launch Information Network Service (ALINS).[7,8]

Notes

[1]Phase 3C will obtain its traditional OSCAR number only after having been successfully placed in operation.
[2]RUDAK is a German acronym for digital repeater. It uses a Mode L frequency pair of 24 cm up, 70 cm down.
[3]Active AMSAT nets are listed on p 64, Oct 1987 *QST*.
[4]The discussion of why stacking is unsatisfactory is beyond the scope of this column. Suffice it to say, I doubt you'll be happy with the results.
[5]M. Davidoff, *Satellite Experimenter's Handbook* (Newington: ARRL, 1984) page 6-3, eq 6.4.
[6]Although a mast-mounted preamp is *desirable* for the 2-meter Mode B downlink, it's *essential* for the 70-cm downlinks of Modes JL, JA and JD.
[7]Information on when ALINS will be on the air will be transmitted on all AMSAT nets and carried on W1AW bulletins as launch day approaches. See note 3.
[8]You can obtain free information on how to get started on Phase 3C—and information about AMSAT in general (including available tracking software)—by sending an SASE to AMSAT, PO Box 27, Washington DC 20044.

Amateur Satellite Communications

Conducted By
Vern "Rip" Riportella, WA2LQQ
PO Box 177, Warwick, NY 10990

Phase 3C: What Do You Need To Work It?

AMSAT's new Phase 3C satellite will be launched soon. As you know from previous columns, the spacecraft carries four transponders: Modes B, JL, S and RUDAK. This month, we'll look briefly at the ground station capabilities required to operate Modes B, JL and S.[1]

Beginning satellite users may be surprised to find that much of the basic ground station equipment is already in the shack or, with a little prudent shopping, can be found at a springtime flea market. Mode B operation on Phase 3C should be the easiest to achieve for the newcomer.[2] To work Mode B, you transmit to the satellite on 70 cm and receive signals from the satellite on 2 meters.[3] Table 1 provides suggestions for the minimum recommended equipment for working Phase 3C, Mode B.

Let's walk through the uplink requirements. The operating frequency is self-explanatory. "EIRP" is Effective Isotropic Radiated Power. You calculate this by adding the gain of your antenna (in dBi) to the power (in dBW) at the antenna feed point.[4] The recommended value of 21.5 dBW corresponds to 141 W EIRP. This can be obtained by feeding 10 W to an antenna having 12 dBic gain.[5] As shown, the polarization should be right-hand circular (RHC). As a compromise, linear polarization, either horizontal or vertical, can be used, but the results obtained may be less satisfactory because of spin modulation.[6]

For your Mode B receive system, the frequency and polarization requirements are self-explanatory. Although the minimum receive antenna gain recommendation is 10 dBic, several commercially available antennas surpass this value by 2 to 3 dB and should be considered. Selecting a higher-gain receive antenna is nearly always worth the cost since it relaxes requirements elsewhere in your system. The exception comes in when you have to elevate the antenna as well as rotate it in azimuth.[7]

The recommendations in Table 1 suggest a preamp is a good idea for Mode B. If you use a short run of good, low-loss transmission line such as Belden 9913, you can leave the preamp in the shack and needn't weatherproof it. Using an expensive preamp with a noise figure of anything less than 2 dB or so is a waste because the sky noise at 145 MHz far exceeds 2 dB. An inexpensive preamp with a gain of 18 to 20 dB should work just fine.[8] I'll talk about effective noise temperature and figure of merit in detail in a future column.

For the experienced satellite user, Mode JL should be the closest thing to perfection yet offered. Mode JL combines 24-cm (1269-MHz) and 2-meter uplinks to

Table 1
Minimum Mode B Station Requirements

Uplink
Frequency: 435.425 to 435.575 MHz.
EIRP: 21.5 dBW for 20 dB peak and 10 dB average SNR on downlink.
Polarization: RHC.
Suitable uplink components: 10 W to a 12-dBic gain antenna.

Downlink
Frequency: 145.975 to 145.825 MHz.
Polarization: RHC.
Minimum recommended antenna gain: 10 dBic.
Maximum receive system effective noise temperature: 625 K (NF = 5 dB).
Minimum figure of merit: −18 dB/K.

Table 2
Minimum Mode JL Station Requirements

Uplink
Frequency: Mode L, 1269.575 to 1269.325 MHz; Mode J: 145.840 MHz ± 20 kHz or 144.450 MHz ± 20 kHz.
EIRP: Mode L, 25 dBW for 20 dB peak and 10 dB average SNR on downlink; Mode J, 25 dBW for 20 dB peak and 10 dB average SNR on downlink.
Polarization: RHC.
Suitable uplink components: Mode L, 10 W to a 15-dBic gain antenna; Mode J, 20 W to 12-dBic gain antenna.

Downlink
Frequency: 435.725 to 435.975 MHz.
Polarization: RHC.
Minimum recommended antenna gain: 13 dBic.
Maximum receive system effective noise temperature: 290 K (NF = 3.0 dB).
Minimum figure of merit: −12 dB/K.

produce a 70-cm downlink. Mode JL's combination of low uplink power requirements, small antennas, low sky noise, low spin modulation, wide passband and so forth, should delight the heart of anyone who has ever wished for a broad-coverage DX satellite. Moreover, the combination of 2-meter and 24-cm uplinks means those living where 24-cm equipment is difficult to obtain can enjoy using 2-meter equipment on this fine mode, just as many enjoyed using Mode J (2 meters up, 70 cm down) on AMSAT OSCAR 8 in the late 1970s. Table 2 presents the minimum recommended station capabilities for Mode JL use. As mentioned earlier, the exact operating frequencies will be published after launch.

Again, the frequency requirements are self-explanatory. The RF power requirements for the uplinks at 24 cm and 2 meters are identical: 25 dBW, which corresponds to 316 watts EIRP. As shown in Table 2, you can get 25 dBW at 24 cm from a 10-W transmitter driving a 15-dBic gain antenna. A small, crossed Yagi or helix quite easily. Many loop Yagis that yield 18 dBi and more are available, but they are linearly polarized. They may be used, although results on Phase 3C remain to be seen. For the 2-meter uplink on Mode JL, a typical setup might include 20 W fed to a 12-dBic gain antenna. If you have more antenna gain, you can roll back the power, and vice versa.

For your Mode JL receive system (see Table 2), good UHF practice is in order. That means a well-designed 70-cm crossed Yagi or helix antenna with RHC polarization. Stay away from those older-design crossed Yagis with the gamma-match feed! Most newer antenna designs use T-matches or folded dipoles with baluns to illuminate the array. Antenna designs have improved greatly in the last few years. You'll get much better results if you use an antenna of recent design. The recommended minimum gain of 13 dBic for your 70-cm receive antenna should be taken as a real minimum. You really don't want to compromise with your receive antenna. If you do, you'll always be wondering why others seem to hear more than you. Besides, some modern designs provide up to 15 dBic of gain in an antenna having a reasonable boom length, and if you invest in a good 70-cm antenna for Mode JL, it will pay off when using it on the Mode B uplink as well.

With Mode JL, you get the chance to field that nice, new GaAsFET preamp, too. On 70 cm, where the sky noise is much lower than on 2 meters, and RFI from terrestrial sources also mitigates, you can use a good, low-noise preamp. One with a noise figure of less than 1 dB is fine. It's a good idea to place the preamp right at the antenna. If, for example, you put 50 feet of transmission line ahead of your preamp and that transmission line has, say, 2 dB loss at 70 cm, you could just as well add that 2 dB to the noise figure of your preamp; the results are practically the same. In other words, you will have worsened the noise figure of your system by 2 dB and wasted much of the benefit you bought the preamp for in the first place.

If you use your 70-cm antenna for receive only (ie, you don't operate Mode

Table 3
Minimum Mode S Station Requirements

Uplink

Frequency: 435.625 MHz, ±15 kHz.
EIRP: Approx 27 dBW under average Mode B AGC conditions
Polarization: RHC
Suitable uplink components: 25 W to a 13-dBic gain antenna.

Downlink

Frequency: 2400.710 MHz, ±15 kHz.
Polarization: RHC.
Minimum recommended antenna gain: 28 dBic.
Typical antenna: 1.4-m dish, assuming 50% efficiency.
Maximum receive system effective noise temperature: 290 K (NF = 3 dB).
Minimum figure of merit: +3 dB/K.

B), you can simply patch your preamp into the transmission line at the antenna in a weatherproof container. If you want to use your 70-cm antenna for both receive and transmit (as most operators do), you'll need to isolate the preamp with transfer relays. In-line preamps with built-in relays are readily available, but cost more. As an alternative, you may opt to find and install your own relays. But placing the preamp at the antenna is most essential if you want to ensure that you'll hear the satellite well.

For the expert satellite user, Mode S offers the newest challenges. For the first time, an Amateur Radio transponder in space will use the S-Band at 2.4 GHz (13 cm).[9] Fortunately, getting on 13 cm these days can be as simple as patching together some basic building blocks. So, even a beginner can succeed if a modicum of care is employed in selecting the system components.

Mode S is designed primarily for a single, narrowband FM signal, but up to four well-spaced SSB stations can also be accommodated. Table 3 provides recommended minimum ground station capabilities for working Mode S.

Because of the high frequencies involved, the Mode S operating frequencies must be viewed as tentative only, and are subject to significant changes when the satellite is finally placed into operation in space. The frequencies shown, however, may be used for general planning purposes. For the 70-cm uplink, one estimate of the required FM uplink power is 27 dBW EIRP, assuming the FM threshold on the downlink is 10 dB. An uplink power of 27 dBW (501 W) EIRP can be generated using 25 W to a 13-dBic gain antenna, for example. If you go up to a 15-dBic antenna similar to one commercial model now available, you can drop the power back 2 dB to 16 W and still get the desired 27 dBW EIRP.

On the downlink side, you'll get to try out your UHF skills. First, a good, low-noise preamp is *absolutely essential*. You simply can't proceed without one *mounted right at your 13-cm receive antenna*. A reasonably priced 13-cm preamp has a noise figure of 1 dB or less. Get the best one you can afford.

This band is just about the lowest frequency where a dish antenna is clearly preferable to a Yagi.[10] A 1.4-m (4.6-ft) dish with a well-designed feed system should provide the minimum recommended gain of 28 dBic. This is about the smallest dish that can be used with obtainable preamp noise figures and this combination will consistently yield the minimum required figure of merit of +3 dB/K. A 1.9-m (6-ft) dish should provide about 30 dBic, and offer a little better FM margin, reduce the preamp requirements, or a little of both.

Notes

[1] I'll cover RUDAK in a later column.
[2] The new Russian satellite RS-10/11 offers an even more modest entry point, requiring only an HF transmitter and receiver to operate one of its modes. See prior installments of this column for details.
[3] The exact frequencies will be published after launch, but approximate values were given in the March column.
[4] Converting from watts to dBW was explained in the March column, and is also addressed in the *ARRL Handbook*.
[5] dBic means decibels referred to an isotropic, circularly polarized source.
[6] Spin modulation results when an elliptically polarized signal is received on a linearly polarized antenna or when a linearly polarized signal is received on an elliptically polarized antenna.
[7] Highly directional antennas are specified for all satellite work. Consequently, the satellite user must accurately point the antennas both in azimuth (compass directions) and elevation. This applies to all modes addressed. Extremely long booms stress the rotators and often complicate installation.
[8] Although your Mode B station will see no benefit from the low-noise performance of a 2-m GaAsFET preamp, the strong-signal overload characteristics of GaAsFETs may warrant their installation for those who operate in dense RF environments, ie, an urban area with numerous repeaters.
[9] Several prior OSCARs have had beacons at 2.4 GHz and above, but Phase 3C is the first to have a *transponder* on this frequency.
[10] Convincing arguments can be made that 13 cm and above is the domain of dish antennas, while the realm of the Yagi and helix is 70 cm and below; for the 24-cm band, it's a toss-up.
[11] Additional information on Phase 3C may be obtained by sending an SASE to AMSAT, PO Box 27, Washington, DC 20044. ▪

Amateur Satellite Communications

Conducted By
Vern "Rip" Riportella, WA2LQQ
PO Box 177, Warwick, NY 10990

Phase 3C Operating Schedules

Last month, I introduced AMSAT's new Phase 3C satellite. Launch is now scheduled for late May, so as you're reading this the launch process could be nearing its countdown stages. Last month, I said there are five key facts you need to know in order to operate the new satellite, and I explained some of them. This month, let's look at another important aspect of operating: operating schedules.

The operating schedule of Phase 3C is directly related to the availability of sunlight to the solar-cell arrays. Without sunlight, the satellite's batteries would soon be exhausted, and the satellite would be forever mute thereafter. Even though the high elliptical orbit of Phase 3C takes it out to 36,000 km (22,000 miles), there are occasions when the positions of the sun, earth and satellite are such that the satellite falls into the earth's shadow. These "eclipses" come and go in cycles. They can occur up to once per orbit, and can last only a few minutes; others stretch to durations of more than an hour.

In any case, the spacecraft cannot be safely operated with a negative power budget (more power being consumed than is being generated) for very long. During an eclipse, the on-board computer[1] and various other subsystems are powered from the battery, but the power-hungry transponders are shut down.

Because the satellite's Mode B receiver and the Mode-JL transmitter both use the 70-cm band, it's impractical to run Modes B and JL simultaneously.[2] Some arrangement needs to be made to tell which mode is in use. This arrangement takes the form of the operating schedule.

The most widely recognized terrestrial "standard" time is Coordinated Universal Time (UTC). When dealing with satellites, however, it's often most convenient to use the time as kept by the satellite as the frame of reference. Let's see how this works and why satellite time is employed in the first place. We'll examine this question by analogy to the earth year and earth day.

A "year," in earth time is simply the length of time it takes the earth to make its way around the sun and return to its place of origin. For smaller time intervals, we use the "day," and then divide it into successively smaller units. Significantly, the length of the day—the time required for the earth to rotate once—is unrelated to the length of the year. That's why the day is about 0.0027378 years (1 year = 365.25 days, approximately), and not some nice even number, say 0.001 (1 year = 1000 days). The rotation of the earth is not, to any degree we can detect, synchronized with the motion of the earth around the sun.

For example, in setting up an operating schedule for the baseball season, we could simply say the regular season runs from day 95 (April 4) to day 265 (September 21). The year's calendar begins with day 1, January 1. That's when we reset the clock (calendar) and start again.

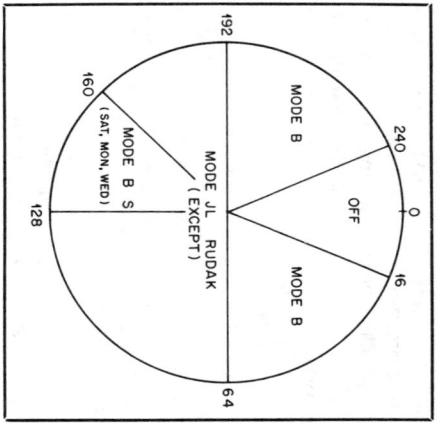

Fig 1—Sample Phase 3C operating schedule.

power, and powers of two are easily handled by computers. So, by "decree" alone, there shall be 256 S-days per S-year. Also by decree, the S-year is said to begin at the instant the satellite passes through its closest point to earth—its perigee. Conversely, halfway through its orbit, when the satellite is at its farthest from earth (apogee), the satellite is at S-day 128, or halfway to 256. Just remember, perigee equals "New Year"!

Now we can express a satellite operating schedule simply in terms of S-days, just as we earlier expressed the baseball season in terms of its start and end dates. We could say, for example, that Phase 3C Mode B will be on from S-day 82 to S-day 150.

In practice, the terms S-day and S-year aren't used, however. These terms are simply constructs for this column. In fact, users of amateur satellites use the term "phase," often designated by the Greek letter phi (φ), or the term "Mean Anomaly" (MA), to refer to the position of the satellite on its orbit (what we've referred to as S-day so far). When MA = 0, the satellite is at S-day 0—perigee. When the satellite is at MA = 128, the satellite is at S-day 128—apogee. Each of the points in between represent equal increments of time.[4]

Let's assume the orbital period is exactly 11 hours. Because the orbit is divided into 256 equal-time increments called MA units or MA ticks, each tick must be 11/256 = 0.043 hour, or 2.58 minutes long. Using our sample operating schedule, when we say Mode B begins at MA 82 and runs to 150, we can now figure out what that means in terms of time of day. Assume perigee occurred at 12:00. Recall that at perigee, MA = 0. To find the time when MA = 82, take 82 × 2.58 = 211.4 minutes = 3.52 hours. So, MA 82 occurs about 3½ hours after perigee, or about 15:31. Now, find when MA 150 occurs. Take 150 × 2.58 = 386.7 minutes = 6.44 hours. So, MA 150 occurs about 6½ hours after perigee, or at about 18:27. We can now see that this Mode B episode lasted from 15:31 to 18:27, or a little less than 3 hours. To double check, take 150 − 82 = 68. Then 68 × 2.58 = 175.44 minutes = 2.92 hours. Is there an easier way to correlate the

schedule for the baseball season, we could simply say the regular season runs from day 95 (April 4) to day 265 (September 21). The year's calendar begins with day 1, January 1. That's when we reset the clock (calendar) and start again.

Let's complete the analogy now. Assume the Phase 3C satellite is the earth. Assume further that its year is defined as the time required for one complete orbit around its "sun," which in this analogy is really the earth. Let's assume that Phase 3C makes one circuit of its elliptical orbit in 11 hours; we could then divide the S-year into S-days (satellite days) of any convenient length. But, just as the rotation of the earth about its axis is unrelated to the movement of the earth around the sun, the rotation of the satellite about its axis is unrelated to its movement around the earth. So, it makes more sense to define an S-day in units that have some significance.[3]

What makes sense is to divide the S-year into 256 S-days. Many of you will recognize 256 as the number 2 raised to the 8th

clock to rise for an early morning orbit!

That would make perfect sense if you had a clock by which to reckon what the MA time was. And, because most earthlings reckon time with clocks that synchronize with the sun (more or less), it sure would be handy to know how to set the alarm clock to rise for an early morning orbit!

operating schedule to time of day? The easiest means by far is to let a computer be your clock. Many AMSAT tracking programs (those of which I'm aware) actually tell you what the exact MA is at that instant. Then, you plan your operating practices according to what the computer's MA clock tells you.

But suppose you don't yet use a computer. How can you tell what the phase or MA is? For this, you will need to know when perigee occurred. This information is available from many sources. AMSAT Nets or AMSAT Area Coordinators can often supply you with this information. When you know when the perigee occurred, you know when the MA clock started, and you can carry on from there.

Finally, Phase 3C will broadcast its own reckoning of the phase or MA in its telemetry. If you can read the PSK or CW telemetry, it will tell you exactly what the MA is. Thereafter, you can synchronize your reckoning with that of the IHU because *it* is the controlling element that determines when to turn on one transponder and turn off another. Thus, the IHU knows where the satellite is in its orbit, and the IHU's reckoning of that position determines what the MA is and what activities and configurations shall occur.

Next month I'll have a feature article on Phase 3C with an overall summary of the spacecraft and what's needed to operate it.[5]

Notes

[1]The on-board computer in the Phase 3 series of spacecraft is called the Integrated Housekeeping Unit, or IHU.

[2]Mode definitions were provided last month in Table 1. Even if there weren't frequency conflict problems with running Modes B and JL concurrently, the combined power consumption of the two would be prohibitively high.

[3]The spin rate (angular velocity) of Phase 3C will be set at about 35 r/min (3.67 radians/sec).

[4]Mean Anomaly, MA, normally refers to the angle that increases uniformly with time to indicate where the satellite is along its orbit. (Compare Mean Anomaly to True Anomaly.) Experts in the field will recognize that MA, as used in this context, clearly refers to a point *in time* rather than an angle. The period of the orbit is divided into equal "time" increments we call MA units. These *do* correlate to positions along the orbit as determined by the angle MA as used in the classical sense. That is, specifying the position in terms of time after perigee is equivalent to specifying the angle that increases uniformly at a known rate beginning at perigee. Thus, the only practical difference in expressing orbital position is the dimension of expression. In one case, degrees are used; in the other, the implied unit is time—perhaps in minutes. We have often referred to ticks of the MA clock, but this is a more ad hoc construct. What may not be clear, however, is that the satellite's velocity changes during the course of its orbit. This can be described in terms of Kepler's second and third laws of planetary motion. So, although MA ticks are uniformly spaced in time, the distance traveled by the satellite between successive ticks varies dramatically.

[5]Free information about Phase 3C and how to get started in the satellite program is available from AMSAT, PO Box 27, Washington, DC 20044.

Amateur Satellite Communications

Conducted By
Vern "Rip" Riportella, WA2LQQ
PO Box 177, Warwick, NY 10990

RUDAK—What Is It?

[Phase 3C operating frequencies: In the May column, I offered the tentative operating frequencies for Phase 3C. My June QST feature article, "Introducing Phase 3C: A New, More Versatile OSCAR,"[1] contains the updated, current assignments.]

In recent months, I've mentioned the Phase 3C RUDAK transponder in this column. Now it's time to explain what it is and the basics of what it does. In this month's column, we're privileged to get a close-up look at RUDAK from information supplied by Hanspeter Kuhlen (DK1YQ), the AMSAT-DL RUDAK Project Manager, and translated by Don Moe (KE6MN/DJ0HC). Next month, I'll have the concluding installment.

RUDAK—Its Beginning

The first RUDAK system design meetings took place early in 1985. This was the time when the packet-radio revolution began spreading around the world, following the acceptance of AX.25 as an international Amateur Radio protocol for digital-link control. It was at these first RUDAK meetings that the payload capabilities of the new AMSAT OSCAR satellite were discussed.

In order to increase the technological challenge for the new satellite project, we decided that a digital packet-radio transponder had to be included. This particular part of the satellite package was named RUDAK, an acronym for *Regenerativer Umsetzer für Digitale Amateurfunk Kommunikation*. In English, this means *Regenerative Transponder for Digital Amateur Communications*. Following another successful ARIANE launch early in March 1988, it was time to get ready for the launch of ARIANE 4 scheduled for June 1988.

RUDAK Experiment Objectives

Providing a reliable digital-communications link for individuals is the main objective of the experiment. Because of the long visibility of the spacecraft in its elliptical (11-plus-hours) orbit, the global footprint of its antenna. The wide range of other objectives can be summarized as follows:

• Real-time digital communications facility with ALOHA[2] (uncoordinated) time-division multiple access (TDMA) and continuous time-division multiplex (TDM) on the downlink. Digipeating with full end-to-end or partial uplink/downlink confirmation. Because of the space-

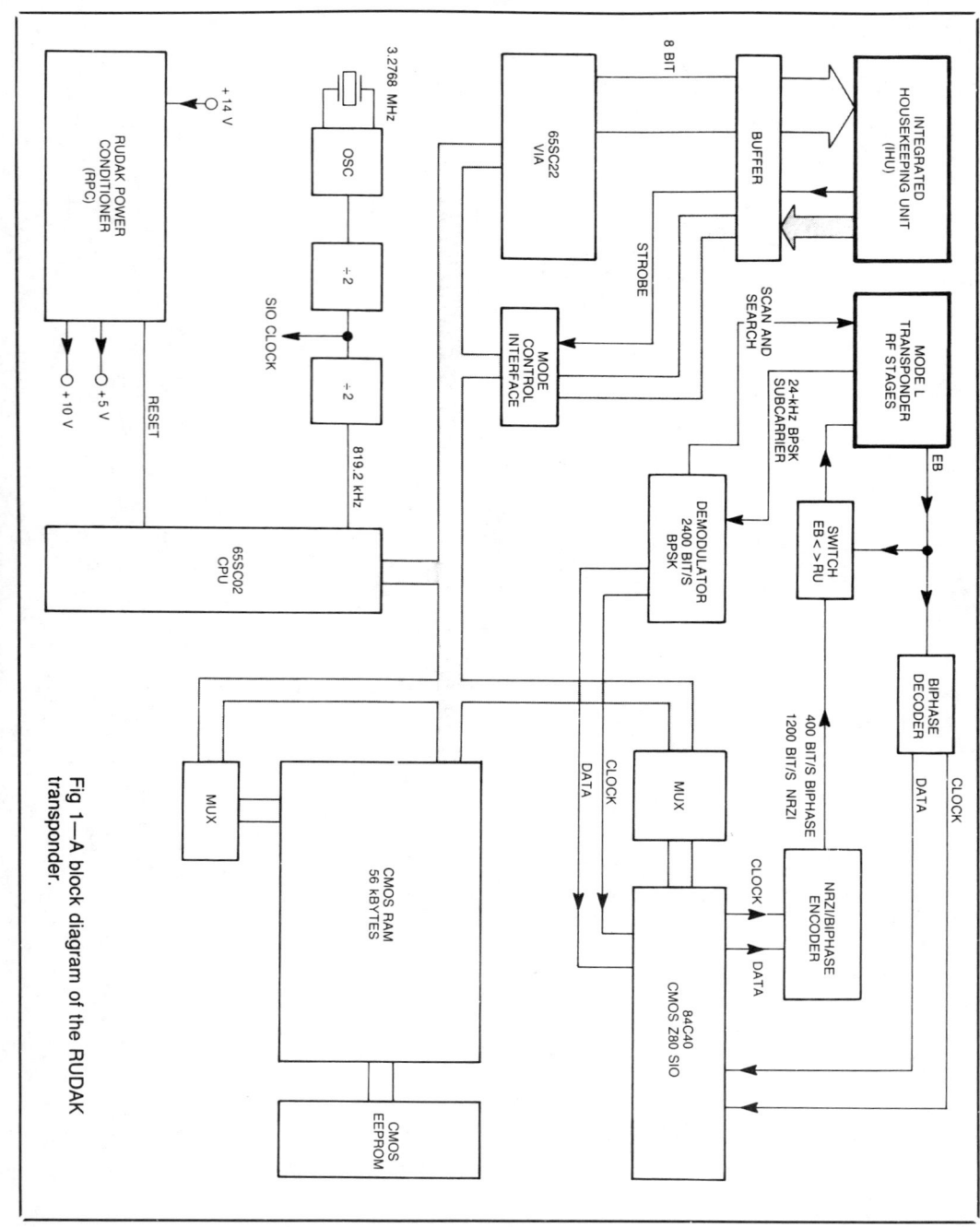

Fig 1—A block diagram of the RUDAK transponder.

craft's extended visibility, a mailbox feature (similar to that provided on Fuji OSCAR 12) was not considered necessary.

• Operational use of binary phase-shift keying (BPSK) with differential coding for the purpose of individual communication, education and experimentation.

• General information broadcast in AX.25 (UI frames) or AMSAT format (ie, 512-byte ASCII).

• Computer-controlled mode switching (autonomous operation).

• An in-orbit facility using a fully programmable computer with intelligent hardware interfaces. This would provide a testbed for communications experiments such as testing alternative access procedures, different bit rates and so forth.

• Keep the size and complexity of ground-station equipment as low as possible to motivate "home-brewing" station equipment.

• Information dissemination through a ROBOT mode to a RUDAK processor. In this mode, ground station connects to RUDAK with the format *call sign and SSID* (secondary-station identifier) are responded to by a transmission of up to 15 different messages. These messages include a valid set of Keplerian elements, present orbit position, general telemetry, an optimum set of AX.25 parameters and so on, followed by a RUDAK disconnect. This feature, in addition to the particular message information, provides a positive confirmation of an established contact with RUDAK.

• Testing of CMOS components (CPU, memory, and so forth) under extreme radiation conditions.

The RUDAK Transponder

A block diagram of the RUDAK transponder is shown in Fig 1. With a regenerative transponder, the uplink signals are first demodulated in a burst demodulator, then processed in a general-purpose communications computer, and eventually modulate a downlink beacon signal (the engineering beacon). The AX.25 protocol handling of the computer is supported by a dedicated interface controller (Z80® SIO) to relieve the computer of unnecessary low-level bit manipulations.

The RUDAK mode is available during mode L operation only. Because the RUDAK transponder is completely independent of the normal passband, however, it will be operative even when the mode L passband is switched off. The frequency plan assigns 1269.710 MHz for the uplink, and 435.677 MHz for the downlink.

Transmissions from ground stations will be of short packets (or bursts) to allow time-sharing of the frequency among several users. A specially developed burst demodulator on board the satellite scans around the center frequency to cope with uncertainties (such as ground-station frequency inaccuracies and Doppler shift) of the uplink signals. A range of ±7.5 kHz is scanned at 120-ms intervals.

The measured performance of the flight unit is about 100 ms for signal acquisition, (ie, the time span between detection of an input signal and its demodulation). The demodulated signal contains two components (recovered data and clock), both of which are available for further processing in the RUDAK computer.

This unavoidable 100-ms overhead has to be considered in setting the appropriate TNC parameter. For instance, the AXDELAY in TNC 1s is set so that a sufficient number of flags ($7E)[3] are transmitted prior to each packet; this allows the demodulator to lock onto the signal. The optimum values for RUDAK communication will be provided via the satellite's beacons. Field tests indicate that the values will probably be not much different from those presently used for the standard FM links.

Because of the inherent collision problem for uplink packets, the theoretically maximum achievable throughput is limited to 18% (ALOHA). In other words, even under optimum channel conditions, only 18% of the packets offered to the channel will survive the competition. This effect is also observable when a mountaintop packet-radio digipeater is within reach of many local stations, but the local stations cannot hear each other. In this case, the carrier-sensing multiple access (CSMA) feature, which normally avoids collisions by inhibiting transmissions on a busy frequency, is useless. If the satellite link did not exhibit a propagation delay of about 300 ms, this problem could be solved with a digipeater operating in the duplex mode. Therefore, for RUDAK, an uplink bit rate of 2400 bit/s has been selected to provide optimum loading of the 400-bit/s downlink. A 1200-bit/s mode is also included on the downlink where NRZI coding is used so as to be fully compatible with the Fuji OSCAR 12 downlink.

The flight model of the RUDAK processor consists solely of CMOS components and requires a total power consumption of 400 mW for 56 kbytes of RAM and 2 kbytes of EEPROM. This allows the software to be RAM resident. Mode L is expected to be the dominating mode throughout the entire Phase 3C mission.

(Stay tuned for more on RUDAK in next month's column.)

Notes

[1]V. Riportella, "Introducing Phase 3C: A New, More Versatile OSCAR," *QST*, Jun 1988, pp 22-30.

[2]For an explanation of ALOHA, see *The 1988 ARRL Handbook*, p 19-40, under the heading "Channel Access."

[3]The dollar sign indicates hexadecimal notation.

Amateur Satellite Communications

Conducted By
Vern "Rip" Riportella, WA2LQQ
PO Box 177, Warwick, NY 10990

The AMSAT RUDAK User Terminals

Last month, I introduced the RUDAK transponder on AMSAT's new Phase 3C satellite (which is now in orbit as OSCAR 13). In that column, the AMSAT-DL RUDAK Program Manager, Hanspeter Kuhlen (DK1YQ), explained what the system did and some of the design rationale. In this month's concluding installment, Hanspeter describes suitable RUDAK user terminals.

For most satellite users, the most interesting part of the RUDAK experiment will be the design and installation of their own satellite terminal. As defined in the objectives (see last month's column), one of the main goals of the RUDAK experiment is to enable reasonably skilled individuals to test modern modes of digital communications. This is reflected not only in the selection and development of an extremely versatile, easy-to-build terminal, but also in the design and development of an extremely versatile, easy-to-build terminal.

Over the course of several months, all the necessary modules for the RF and digital unit were designed and tested. Wherever possible, we used off-the-shelf designs in order to avoid reinventing the wheel. PC boards are now available for these modules, and have been successfully beta-tested by several amateurs.

The RUDAK User Terminal consists of two separate sections: the RF and Digital units. Their features can be summarized as follows:

- Operation in *all* satellite modes: CW, SSB, PSK through passband, Fuji-OSCAR 12, reception of UoSAT 1 and 2 bulletins and, of course, RUDAK.
- The RF unit can be used as a general-purpose power amplifier for terrestrial 23-cm communications.
- Using a hybrid PA module, the power amplifier provides 20 W CW output on 24 cm. In combination with a 15 dBi-gain antenna, this power output level is sufficient for reliable data communication via RUDAK and the other satellite modes.
- Conversion of 2-meter signals to 24 cm; a built-in attenuator is capable of accepting 1 W of driving power.
- A 2400 bit/s modulator.
- BPSK demodulators for 400 and 1200 bit/s with biphase and NRZI coding, respectively.
- Compatibility with Fuji-OSCAR 12 formats.
- Built-in TNC.
- Internal switching for space or terrestrial packet-radio operation.
- AMSAT interface to general-purpose computer (Atari® 800XL) for satellite tracking that features automated antenna azimuth and elevation control, satellite telemetry decoding and display of measured engineering-unit parameters, visibility prediction, data communications in AMSAT block format similar to the telemetry blocks of OSCAR 10, and so on. (To the best of our knowledge, the Atari 800XL computer is available in almost any country in the world. This computer was selected for OSCAR 10 satellite-control purposes because of its extremely low price and—even more important [sometimes]—because of its very effective RF shielding. As soon as another computer is operated close to the sensitive satellite receive antenna, this shielding is appreciated even more.)
- Common power supply in the RF unit.

For the digital unit, we selected a so-called modular design. Commencing from a basic version, a station can be expanded step-by-step to include additional functions. The whole setup is shown in

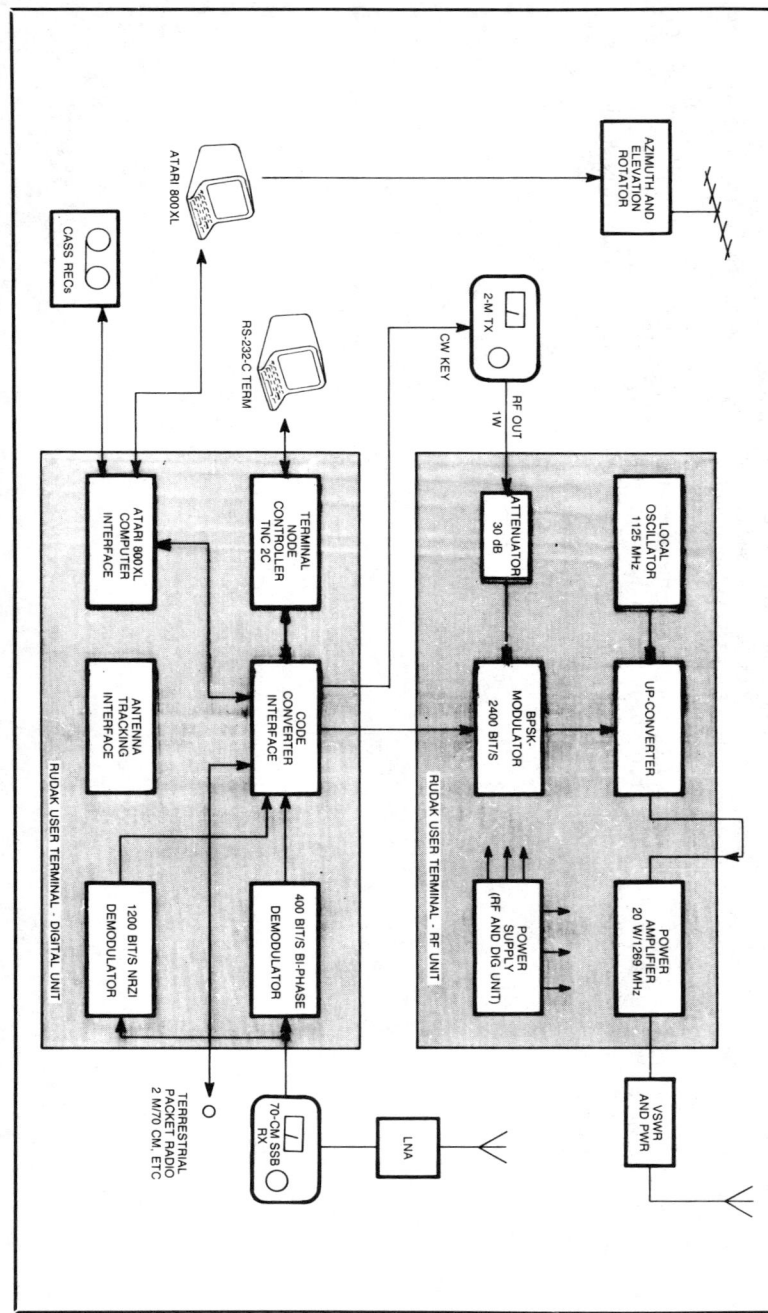

Fig 2—Block diagram of a complete (super-duper) RUDAK user station. Don't let the complexity scare you; you can start off observing RUDAK operations with much less equipment (see text).

Satellite Anthology 84

Fig 2. This block diagram shows the equipment required in addition to the terminal.

Before becoming concerned by the relatively complex configuration shown in Fig 2, remember that it shows the ultimate—the super-duper—version of an amateur satellite home station. In addition to a receiver and an RS-232-C terminal, the minimum required hardware includes a 400 bit/s demodulator, the code converter and a TNC. This allows you to monitor RUDAK activities as an observer. Such observation (hopefully coupled with hearing a lot of exotic call signs) should definitely motivate you to do more and eventually air your own signal.

Great care and attention has been devoted to keeping the station easy to build, eliminating the need for special tools or machinery. Several components, particularly those in the RF unit, have been built and evaluated in many different design approaches, testing their suitability for amateur construction. The results of these investigations have been compiled in the RUDAK User Manual. This manual provides the necessary background information on all the experimental aspects as well as a detailed description of how to build your own terminal.

The RUDAK Field Test

Because the RUDAK experiment uses several newly developed items (both in the hardware and software areas), a comprehensive field test for all components was mandatory. During this test, the equipment was installed atop a 45-meter-high (148-foot) water tower in Ismaning (near Munich) and provided a test-bed for several amateurs in the Munich area.

The test-bed RUDAK was used exactly as we expected it would be used while in orbit aboard AMSAT Phase 3C. The operation included testing of the satellite main computer (IHU) and the internal communications between the two computers. The equipment configuration of the field test emulates the on-board equipment so that any hardware or other improvements can also be incorporated into the flight package.

Acknowledgments

Many thanks to Dr Karl Meinzer (DJ4ZC) and Werner Haas (DJ5KQ) for their generous support. Thanks also to the fantastic people on the RUDAK team for their excellent and invaluable contributions: Peter Gülzow (DB2OS), Stefan Eckardt (DL2MDL), Gerhard Metz (DG2CV), Knut Brenndörfer (DF8CA), Herrmann Hagn (DK8CI), Don Moe (KE6MN/DJ0HC) of AMSAT-NA and Robin Gape (G8DQX) of AMSAT-UK.

As you may infer from the foregoing, using the new RUDAK should be challenging and enjoyable, but not overwhelmingly difficult. Our thanks to DK1YQ and his RUDAK team for sharing their project with us. The field of digital Amateur Radio satellite transponders now includes UoSAT-OSCAR 11 (developmental use only), Fuji-OSCAR 12 (Mode JD) and the new AMSAT-OSCAR 13 (formerly Phase 3C).

During 1989, however, the field will truly blossom with perhaps as many as a half dozen new OSCARs of a brand new class of satellites being pioneered by AMSAT-NA. The existence of this new class of OSCAR will do several things. First, it will reassert Amateur Radio in general—and the Amateur Radio space program in particular—as a place where innovative things continue to occur not by happenstance, but by design. This is at once in the best tradition of our hobby, and simultaneously expands the envelope of what we're about. Second, it emphasizes and justifies our collective occupancy of the valuable spectrum (which is, after all, our life's blood) on the basis that we actually *do* advance the state of the art to society's benefit. And third, the new class of OSCARs will provide a basis for unprecedented networking among radio amateurs. Thus, it would be well to take spade in hand *now* and prepare *your* spacecomm system for the imminent verdant spring. By this time next year, you may be surprised to find yourself enveloped in a virtual flowering field of new OSCARs spawning possibilities stretching even fertile imaginations.

Stay tuned! Next month, first reports on operating the new OSCAR.

Amateur Satellite Communications

Conducted By
Vern "Rip" Riportella, WA2LQQ
PO Box 177, Warwick, NY 10990

Birth of a New OSCAR: First All-Japanese Project Debuts

The Japanese Amateur Space Program has produced its first successful all-Japanese satellite with the flawless launch and initial operation of Fuji-OSCAR 12. The new satellite was known as JAS-1 prior to launch. A joint project of the Japan Amateur Radio League (JARL), Japan AMSAT (JAMSAT), Nippon Electric Company (NEC) and the Japanese National Space Agency (NASDA), the newest OSCAR is the result of a half decade of planning. The overall specifications of FO-12 were provided in this column in June 1986. Now we report on the launch and initial operation of FO-12. Just locating FO-12 after launch turned into one of the most intriguing episodes in memory!

The launch from southern Japan occurred at 20:45 UTC August 12. At precisely 21:47:07 UTC, FO-12 was born. At the exact moment of deployment from the launcher, the FO-12 435.795-MHz beacon was activated as planned over South America. It was immediately heard by CE3GA at Santiago University and by Junior DeCastro, PY2BIO, in Sao Paulo. Twenty minutes later, amateurs in Europe got their first look at FO-12.

The AMSAT Launch Information Network Service (ALINS), organized for the FO-12 launch, was by all accounts the best ever provided for an OSCAR launch. A hybrid network of radio and telephone links established a network spanning the entire globe. Few, if any, areas were left without a live network feed providing information on the progress of the launch as it happened. JA1ANG in Tokyo provided countdown coverage as the H1 vehicle was launched, and he described progress of the launch. CE3GA in Santiago provided live pickup of the first FO-12 telemetry signals, which were relayed to W3IWI on 15 meters and then to the network. G3YJO provided similar telemetry coverage when FO-12 appeared over London. WØRPK provided continuity and background information as network control from Iowa. Transmitting the live telephone ALINS audio on the HF bands were WA3NAN, WA2LQQ, W6SP, WØRPK, W6GC, ZS6AKV and W3IWI with a bi-directional link on 21.390 to South America.

Initial reports on Mode JA indicated transponder performance was excellent. The receiver was superbly sensitive. NK6K reported satisfactory results with 1 watt to his uplink antenna. He suggested FO-12 could probably be worked with an "omni" if one cared to use a bit more RF at the antenna. Most transponder users reported very deep fades on the downlink. This is due to the random spinning of the satellite in its early life. The tumbling will dampen out in a few weeks due to the passive stabilization used into FO-12. Like previous OSCARs, FO-12 has a simple bar magnet to dampen tumbling. The interaction with the geomagnetic field produces a small but persistent torque which reduces the tumbling motion of the spacecraft.

Initial tracking of FO-12 immediately after deployment was accomplished simply and accurately by adjusting the prelaunch estimates to account for the 14-minute delay of the launch from 2031 to 2045 UTC on August 12. Using these adjusted prelaunch estimates, amateurs tracked FO-12 easily.

Curiously, however, when NASA's first tracking data for FO-12 became available and was loaded into computerized satellite-tracking programs, the results were disappointing: they were off target by an uncharacteristically large degree. The adjusted prelaunch values continued to be the best available numbers until August 15. By then, it began to appear that NASA had mislabeled the satellites launched by the Japanese H1 vehicle. NASA was calling object 16908 the Experimental Geodetic Payload (EGP); 16909 was called JAS-1 and 16910 was labeled the rocket body. But there was accumulating evidence that 16908 might be JAS-1 (FO-12) and not EGP! This evidence included numerous observations by dozens of experienced satellite trackers, including NK6K, WØRPK, WA3WBU, W2RS and KA9Q.

In 1978, AMSAT-OSCAR 8 was confused with another object launched with it. AMSAT helped sort out several of the ta's of Aug 19, 1986

Table 1
Revised Orbital Elements†

Reference Epoch	86 230.26525652
Element Set	6
Inclination	50.0097
Right Ascension of Ascending Node (RAAN)	236.4938
Eccentricity	0.0011125
Argument of Perigee	233.7423
Mean Anomaly	353.0984
Mean Motion	12.44393428
Revolution	67
Drag (Decay rate)	-3.9e-07

†as of Aug 19, 1986

Soviet RS birds that were launched in a flock of six (RS-3 through 8).

A possible answer to the puzzle of which object was really FO-12 came August 15. It came in two parts. First, a NASA employee at the Goddard Space Flight Center explained how the satellites get labeled. Normally, North American Air Defense (NORAD) labels the new spacecraft in the order in which they are observed. Thus, 16908 would have been the first observed by NORAD's radar. NORAD knew from NASDA, the Japanese launch authority, the first object to be deployed was EGP. So NORAD expected the first object spotted would be EGP. Thus, EGP became the name of the first object NORAD actually observed in the group. Was it really EGP?

A second revelation came when Miki, JR1SWB, pointed out that JAS-1 was deployed downward or backwards. This meant it would have had a slightly lower altitude and thus a slightly higher mean motion than EGP and the rocket's second stage. Since JAS-1 had a higher mean motion, it had a higher angular velocity. Had it shown up first on NORAD radar even though it was deployed after EGP? It seemed true since NASA's revised data for object 16908 soon was found to fit almost exactly with precise radio observations of FO-12.

Then, on August 16, W3IWI reported that the EGP experimenters were getting excellent laser ranging from EGP and had fixed its time and position precisely. (EGP is loaded with mirrors and laser retro-reflectors specifically designed for this type of ranging and orbit determination.) The next event was a bombshell for the earlier hypothesis that FO-12 was really EGP and not FO-12 after all?

But how could the excellent fit of 16908 with the radio observations of FO-12 be accounted for? Objects 16909 and 16910 were thousands of miles away, so neither of them could possibly be FO-12. The dilemma was essentially this. Everyone knew precisely where 16908 was. But the EGP people had excellent data to show 16908 was actually EGP. On the other hand, AMSAT had excellent data to show 16908 was FO-12! How could both be correct?

KA9Q and this conductor concluded the most reasonable explanation that fit the observations was that 16908 was actually EGP after all, but that neither 16909 nor 16910 was FO-12. Furthermore, FO-12 had to be in the close vicinity of EGP. Perhaps it just hadn't been "seen" yet. Maybe it had not yet been cataloged by NORAD or NASA. To test this hypothesis, KA9Q ran another precise Doppler curve to detect TCA, the exact Time of Closest Approach. He calculated FO-12 was leading 16908 by 5 seconds at 08:37 UTC on 17 August. It was then that the big light bulb went on in several heads.

In W3IWI's EGP report, he mentioned the EGP experimenters had visually observed EGP quite easily. Significantly, they noted a small object leading EGP by about 3 seconds. Could it have been FO-12? The data seemed to suggest so. For the present, AMSAT believes object 16908 is in fact EGP by perhaps 5 seconds or about 20 miles and diverging slowly from it. Stay tuned for further developments in the saga!

Amateurs have reported spectacular visual

observations associated with the launch of Japan's H1 vehicle. At least one observer, VE3GSO in London, Ontario, reports having seen all three major objects launched, the EGP FO-12 and the rocket body. The rocket body was reported to have been enveloped in a bluish, iridescent cloud of gas. Similar reports made news headlines across North America. The gas was apparently ionized hydrogen released by the intentional venting of the launcher's tanks and the subsequent effect of solar radiation. The solar radiation ionized the hydrogen much as it does atmospheric hydrogen.

VE3GSO says he actually saw FO-12 with binoculars. It appeared bluish with a slight twinkle. Sources indicate the blue color results from the solar cells and their coatings. EGP is much more easily visible. Covered with dozens of laser retro-reflectors and plane mirrors, it is specifically designed to be ranged by laser and visually spotted. Its primary objective is to provide precise information on the position and movements of various observation sites on the earth. About 7 feet in diameter, the EGP can be as bright as a first-magnitude star when observed under optimum lighting conditions. According to W3IWI, its brightness can vary from 1 to 4 while most of the time being around 2, the apparent magnitude of Polaris, the North Star.

EGP should be visible between 15 minutes after sunset and through perhaps local midnight according to KA9Q. Similarly, it should be visible from up to four hours before dawn through about 15 minutes before dawn. According to VE3GSO, there was a prominent twinkle to EGP when he saw it due, he says, to the effects of the many mirrors on the large sphere. EGP was the primary payload on the H1 launch.

Next month, we'll take a tour of basic FO-12 operating procedures.

Amateur Satellite Communications

Introducing Japanese Amateur Satellite Number One (JAS-1)

In nature, a sign of a healthy, prosperous species is often proliferation. It's true in Amateur Radio satellites as well. The latest to appear on the scene is JAS-1. The specifications below were provided by Tak Okamoto, N6MBM/JE2PKI, and Harold Price, NK6K.

JAS-1 is a joint effort of many organizations. Besides JARL (Japan Amateur Radio League) and NASDA (Japanese national space agency), the Nippon Electric Company (NEC) built "system" units (space frame, power supply etc), JAMSAT (Japan AMSAT) designed and built the "mission" units (transponders, telemetry/command and housekeeping microcomputer) and ground-support systems.

JAS-1 Mission Objectives:

- Provide reliable worldwide Amateur Radio communications.
- Enable radio amateurs to study tracking and command techniques.
- Offer an in-space "proving ground" for radio amateur developed and built transponders and subsystems.
- Provide NASDA an opportunity to carry out a "multipayload" launch using their new "H-1" launcher. (NASDA has never engaged in a multipayload launch, thus the JAS-1 project will offer NASDA an excellent opportunity by providing them with an active payload having its own telemetry-beacon and transponder for ranging.)

Form and General Dimensions: The spacecraft is a 26-facet polyhedron, which measures 400 mm × 400 mm × 470 mm (15.75 in × 15.75 in × 18.5 in) and weighs 50 kg (110.2 lbs).

Launch and Orbit: JAS-1 will be launched into a circular low-earth orbit, which will be non-sun synchronous and non-polar.

Launch vehicle: H-1 2-stage rocket
Launch number: Test Flight 1
Launch site: Tanegashima Island, Japan
Launch date: August 1986
Estimated inclination: 50 degrees
Estimated altitude: 1500 km
Estimated period: 120 minutes
Estimated window per pass: 20 minutes/pass
Estimated passes per day: 8 passes/day

Designed Life: Estimated lifetime is three years.

Special Features of JAS-1: JAS-1 will carry two separate Mode J (2-meter uplink, 70-cm downlink) transponders. One is a linear transponder, and the other is a digital "store-and-forward" transponder mainly for non-real-time communication between stations located in different time zones. The digital transponder will provide "error-free" information exchange.

Transponders:

a) The linear transponder: Mode JA

The passband is 100 kHz wide. The transponder has an output of 1-W PEP. Ground stations will need an uplink power of 100-W EIRP. The sidebands are reversed, ie, the uplink is LSB and the downlink is USB. There

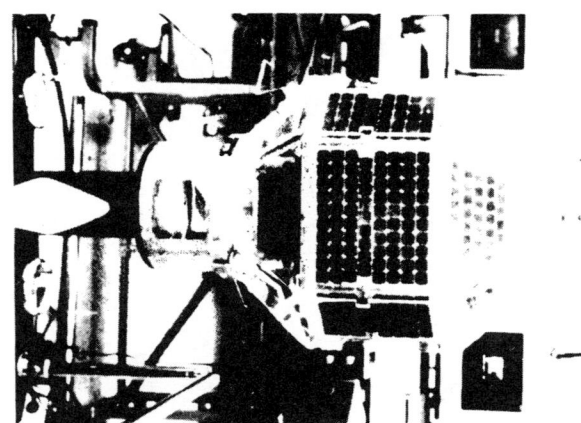

is a 100-mW CW beacon switchable to PSK when needed.

Uplink passband: 145.90 MHz–146.00 MHz.
Downlink passband: 435.80 MHz–435.90 MHz.
Beacon frequency: 435.795 MHz.
Translate frequency: 581.80 MHz.

b) The digital transponder: Mode JD.

There are four 145-MHz-band input channels using Manchester-coded FM for the uplink. Ground stations will need 100-W EIRP. There is one downlink channel in the 435-MHz band using PSK; the output is 1-W RMS. Channels are:

Uplink channel 1: 145.850 MHz.
Uplink channel 2: 145.870 MHz.
Uplink channel 3: 145.890 MHz.
Uplink channel 4: 145.910 MHz.
Downlink channel: 435.910 MHz.

The data format is HDLC. The protocol is AX.25 Level 2 Version 2. The data transfer rate is 1200 bit/s for both uplink and downlink.

JAS-1 will be a store-and-forward system but not a real-time digipeater. Digipeating is ineffective in low orbit.

JAS-1 has four uplink channels for one downlink channel. This is attributable to the differences in channel efficiency between uplink and downlink. An uplink channel will be shared by several ground users. Since the ground users won't hear each other, the uplinks will be subject to packet collisions. This "Pure ALOHA" system has maximum channel throughput of 18.4%. But the JAS-1 downlink will be 100% efficient since only JAS-1 transmits there. To balance capacity and add redundancy, four uplink channels are used. The combined uplink efficiency will

then be 4 × 18.4% or 73.6%. The remaining downlink time will be used for general messages and telemetry data.

Digital Hardware: The microprocessor is a MIL-STD-883B screened NSC-800. It controls the digital transponder and is the IHU (Integrated Housekeeping Unit). The memory is 1.5 megabyte (Mbyte). Forty-eight 256-kbyte NMOS DRAMs are used. A hardware-based error-detection/correction circuit is incorporated to protect the entire 1.5 Mbyte and provide a 1-Mbyte error-free memory area. The rest will be used for message storage. The memory unit is physically divided into four identical 256-kbyte memory cards, any one of which can be assigned as the system area. Up to three cards can be turned off. This design provides system redundancy and allows command stations to control power consumption without a total loss of service.

JAS-1 has five hardware HDLC controllers: Four of them are for the uplink channels and one is for the downlink channel.

Power System: Twenty-five of JAS-1's 26 faces are covered with a total of 979 solar cells and will initially generate 8.5 W. JAS-1 employs 11 NiCd battery cells with a capacity oi 6 Ah. These supply an average 14 V to the JAS-1 main power bus. The 14 V is converted and regulated to +10 V, +5 V and −5 V.

Antenna System: JAS-1 has three antennas.

2-m receive antenna: slant ¼-wave monopole; isotropic; −4 dBi gain 70-cm transmission antenna.

Mode-JA: slant turnstile LHCP + Z axis + 3 dBi gain.

Mode-JD: slant turnstile RHCP − Z axis + 3 dBi gain.

Attitude Control: Forced shaking using the earth's geomagnetic field. JAS-1 has two 1 TAm² permanent magnets in its Z axis.

Telemetry: The analog system telemetry has 12 analog channels and 33 system status flags. The telemetry is sent on the 100-mW beacon on 435.795 MHz in CW, switchable to PSK. The digital system telemetry has 29 analog channels and 33 system status flags. This software-driven telemetry can be sent in any format and can include short text messages. This telemetry can be sent on either the Mode JD downlink channel (435.910 MHz) or the Mode JA CW beacon (435.795 MHz).

Command: A simple three-channel telecommand system is used for global control functions. An additional 37 channels are available, mainly for controlling the digital transponder. On-board command from the NSC-800 is also available.

Ground Stations:

Mode-JA: A station with a 10-W 2-m SSB transmitter and a 10-dBi beam for uplink will suffice. A 70-cm receiver (with low NF) with a 15-dBi beam for downlink should be adequate.

Mode-JD: In addition to the Mode-JA

setup, FM mode will be required for the 2-m transmitter.

Since JAS-1 uses the standard AX.25 protocol and 1200-bit/s data rate, ground stations will be able to use a TAPR-style TNC, a 2-m FM transmitter and a 70-cm receiver without modification.

The JAS-1 modem and a special interface board will be made available. [Additional information and a schematic diagram of the JAS-1 modem circuit appear in the *5th Computer Networking Conference* publication (see "Outline of Satellite JAS-1," by Fujio Yamashita, JS1UKR). See page 140, this issue, for ordering information. Ed.] It will contain the Manchester modulator and an audio PSK demodulator, allowing connection to the "modem disconnect" connector of a TAPR-style TNC. The modem also connects to the audio input and PTT of the 2-m FM transmitter and to the audio output and frequency control (option) of a 70-cm SSB receiver. Although JAS-1 will be available to individual access, the general amateur community will benefit from "JAS-1 gateways." Messages relayed through gateways can be sent worldwide and will be as easy as sending messages to distant stations via a WØRLI HF gateway.

Watch for word on the launch of JAS-1 this August on AMSAT nets and on ARRL bulletins from W1AW.[1-3]

[This column will not appear in July and August, but will return in September.—Ed.]

Notes

[1] A list of active AMSAT nets is available from the author for a business-sized SASE.

[2] Information about getting started on OSCAR, the AMSAT Software Exchange and AMSAT membership can be obtained for a business-sized SASE to the author.

[3] Help in getting on OSCAR may be as close as your nearest AMSAT Area Coordinator. For his name and location, call Jack Somers, WA6VGS, at 800-421-6621.

Amateur Satellite Communications

Conducted By
Vern "Rip" Riportella, WA2LQQ
PO Box 177, Warwick, NY 10990

Operating the Flying Mailbox: FO-12 Mode JD

In the June 1986 column, we introduced JAS-1. With its successful launch on August 12 and the commencement of operations shortly thereafter, JAS-1 has become Fuji-OSCAR 12, or FO-12.

Last month, we traced the excitement accompanying FO-12's birth. Literally thousands of QSOs have since occurred on FO-12's Mode JA transponder. It's a linear transponder best suited for SSB and CW QSOs. Soon, however, FO-12's Mode JD digital transponder will be placed on line.

This month, we'll look at how to operate Mode JD's electronic mailbox. The following was prepared by Tom Clark, W3IWI, to whom we are indebted. Frequencies for Mode JD were provided in the June column.

FO-12 Telemetry Data Format

JAS-1 FF YY/MM/DD HH:MM:SS
xxx xxx xxx xxx xxx xxx xxx xxx
xxx xxx xxx xxx xxx xxx xxx xxx
xxx xxx xxx xxx xxx xxx xxx xxx
xxx xxx xxx xxx xxx xxx xxx xxx
sss sss sss sss sss sss sss sss
sss sss sss sss sss sss sss sss
sss sss sss sss sss sss sss sss
yyy yyy yyy yyy yyy yyy yyy yyy

FF: = Frame Identifier RA: Realtime Telemetry (ASCII)
RB: Realtime Telemetry (Binary)
SA: Stored Telemetry (ASCII)
SB: Stored Telemetry (Binary)
M0: Message -0
M1: Message -1
M9: Message -9

YY/MM/DD = Date
HH:MM:SS = Time (UTC)
(Following is valid only for RA and SA frames)
xxx = 000-999 Format: 3 digit decimal (Analog Data) 27 samples in row 0 column 0 through row 2 column 6 (denoted #00-#26 below)
y = 0-F One byte Hex (System Status Data) 9 samples in row 2 column 7 through row 2 column 9 (denoted #27a-#29c below)
s = 0 or 1 Binary Status Data 30 samples in row 3 through row 3 column 9 (denoted #30a-#39c below)

FO-12 Telemetry Calibration Equations

Channel	Item	Equation	Units
#00	Total Solar Array Current	1.91 * (N − 4)	mA
#01	Battery Charge/Discharge	3.81 * (N − 264)	mA
#02	Battery Voltage	N * 0.0210	V
#03	Half-Battery Voltage	N * 0.00937	V
#04	Bus Voltage	N * 0.0192	V
#05	+5 V Regulator Voltage	N * −0.00572	V
#06	−5 V Regulator Voltage	N * −0.00572	V
#07	+10 V Regulator Voltage	N * 0.0116	V
#08	JTA Power Output	5.1 * (N − 158)	mW
#09	JTD Power Output	5.4 * (N − 116)	mW
#10	Calibration Voltage #1	N / 500	V
#11	Offset Voltage #1	N / 500	V
#12	Battery Temperature	0.139 * (689 − N)	Deg C
#13	Baseplate Temperature #1	0.139 * (689 − N)	Deg C
#14	Baseplate Temperature #2	0.139 * (689 − N)	Deg C
#15	Baseplate Temperature #3	0.139 * (689 − N)	Deg C
#16	Baseplate Temperature #4	0.139 * (689 − N)	Deg C
#17	Baseplate Temperature #5	0.139 * (689 − N)	Deg C
#18	Temperature Calibration #1	N / 500	V
#19	Offset Voltage #2	N / 500	V
#20	Facet Temperature #1	0.38 * (N − 684)	Deg C
#21	Facet Temperature #2	0.38 * (N − 684)	Deg C
#22	Facet Temperature #3	0.38 * (N − 690)	Deg C
#23	Facet Temperature #4	0.38 * (N − 683)	Deg C
#24	Facet Temperature #5	0.38 * (N − 689)	Deg C
#25	Temperature Calibration #2	N / 500	V
#26	Temperature Calibration #3	N / 500	V

FO-12 System Status Telemetry Bytes

Channel	Item
#27a	Spare (TBD)
#27b	Spare (TBD)
#27c	Spare (TBD)
#28a	Spare (TBD)
#28b	Spare (TBD)
#28c	Memory Unit #0 error count
#29a	Memory Unit #1 error count
#29b	Memory Unit #2 error count
#29c	Memory Unit #3 error count

FO-12 Binary Status Data Points

Channel	Item	1	0
#30a	JTA Power	On	Off
#30b	JTD Power	On	Off
#30c	JTA Beacon	PSK	CW
#31a	UVC Status	On	Off
#31b	UVC Level	1	2
#31c	Main Relay	On	Off
#32a	Engineering Data #1	Full	—
#32b	Battery Status	Tric	Full
#32c	Engineering Logic	Tric	—
#33a	Engineering Data #2	On	Off
#33b	PCU Status Bit 1 (LSB)	On	Off
#33c	PCU Status Bit 2 (MSB)	On	Off
#34a	Memory Unit #0	On	Off
#34b	Memory Unit #1	On	Off
#34c	Memory Unit #2	On	Off
#35a	Memory Unit #3	On	Off
#35b	Memory Select Bit 1 (LSB)	—	—
#35c	Memory Select Bit 2 (MSB)	—	—
#36a	Engineering Data #3	—	—
#36b	Engineering Data #4	—	—
#37a	Computer Power	On	Off
#37b	Engineering Data #5	—	—
#38a	Solar Panel #1	Lit	Dark
#38b	Solar Panel #2	Lit	Dark
#39a	Solar Panel #3	Lit	Dark
#39b	Solar Panel #4	Lit	Dark
#39c	Solar Panel #5	Lit	Dark
	Engineering Data #6	CPU	TLM
	CW Beacon Source	—	—
	Engineering Data #7	—	—

Example:

FO-12 RA 86/08/01 09:00:00
500 xxxxxx xxx xxx xxx xxx xxx xxx
xxx xxxxxx xxx xxx xxx xxx xxx xxx
xxx xxxxxx xxx xxx xxx 000 004 yyy.
01s sss sss sss sss sss sss sss sss

Real time ASCII frame sent on 86/08/01 at 09:00:00 UTC

Total Solar Array Current = 947 mA
Memory Unit #0 error count = 4
JTA power Off
JTD power On

FO-12 Packet BBS User Interface Information

Mailbox Commands (Basic users training)
(WØRLI/WA7MBL equivalences added by W3IWI)

1. Summary

1.1 Available Commands

F: List files addressed to ALL or to current user
H: Help
K: Kill file(s)
M: List file(s) to/from current user
R: Read file(s)
W: Write file

1.2 Command Syntax

The general format is: <a command letter> <space> <argument>. At least one blank is required between <a command letter> and <argument>.

2. Command Prompt

FO-12 Mailbox supplies a prompt "JAS>" with no CR nor LF to indicate that the system is ready to accept a command from the user. A user can "type ahead" commands while FO-12 is sending messages or data to the user. FO-12 will execute the commands in the waiting queue later.

3. Commands

3.1 The "F" Command

F = FILES. Shows the latest 10 files the first time it is entered during a session. Subsequent 'F' commands will list the next 10 active files (messages). A message posted to multiple users has "*" in its 'To:' destination field. See also the "M" command described below. (The WØRLI/WA7MBL equivalent command is LL 10 the first time you send an "F".)

Example:

```
JAS>F
NO. DATE FROM TO SUBJECT
117 10/12 F8ZS ALL ARSENE update
116 10/12 DL3AH ALL Abgeleichanleitung
115 10/12 W3KH ALL Dish Design Specs
114 10/11 JAIRL ALL JAS-1 new schedule
113 10/11 WA2LQQ ALL ALINS for Phase-3C
112 10/10 JA1DSI ALL WHO MANAGES HK0XXX QSL?
```

111	10/10	G3AAJ	*	Harry in London
110	10/09	W0RPK	ALL	P-3C countdown #8
107	10/09	9M2CR	ALL	NMCR AMTOR mailbox now QRV
103	10/06	JR1FIG	JA9BOH	Uchiawase wa raishuu?
102	10/09	N7FDA	*	RS-232C card for PC-108

JAS>F

101	10/09	G3RUH	ALL	New software for BBC
100	10/08	JR1ING	JR1FIG	Sara ni kogata no TNC
99	10/08	JA1TUR	ALL	AFDEM-JA #3 in progress
98	10/08	N5AHD	ALL	Call for papers
96	10/08	KA9Q	ALL	TCP/IP on TAPR NNC
95	10/08	N5AHD	JR1FIG	Automatic tracking system
94	10/07	DJ5KQ	ALL	IPS-RA enhancements
93	10/07	DB2OS	ALL	Wettersatelliten
92	10/07	DB2OS	ALL	RUDAK-Statusreport
85	10/07	5H3KK	ALL	Now QRV on FO-12

3.2 The "R" Command

R <file#1>, <file#2>, <file#3> . . . <file#7>, <file#8>
R = READ. Read file(s) (messages) specified by file number(s) you got from the "F" command. Up to eight files can be specified. (The W0RLI/WA7MBL equivalent command is also "R" except that you may specify multiple files to be read on FO-12.)

Example:

JAS>R 95,102
Posted: 86/10/08 17:33 UTC
From: N5AHD
To: JR1FIG
Subj: Automatic tracking system
Dear Saya,
Thank you for the compliments on the manual you received from G3AAJ. Two computers are now used—one for control of antenna system, radios [etc]

Posted: 86/10/09 03:21:42 UTC
From: N7FDA
To: JR1FIG,JA1JHF
Subj: RS-232C card for PC-1089

Saya, I need one more RS-232C card for my old faithful PC-1089. Would you ask Kanawa san if he could still get one in Akihabara? Miki

3.3 The "W" Command

W [call1, call2, call3 . . . call7, call8]
W = Write. Send a message (file) to others. As many as eight destination addresses can be specified. The part of the command line in brackets [call1, call2, call3...] is optional. A message without specific destination is "public," ie, addressed to "ALL."

The JAS-1 mailbox will then prompt you to send the subject field by sending "Subj:". You can send a subject field with up to a 32 character string. After receiving the "Text:" prompt, you enter the message text, ending each line with <cr> (carriage return). You terminate with either a

<cr> . <cr>   or
<cr> <ctl-Z> <cr>

(ie, a line containing only a period or a control-Z) to indicate the end of your text. (The W0RLI/WA7MBL equivalent command is "S" except that multiple addresses can be used. Entering only W is equivalent to S ALL)

Example:
JAS>W N7FDA Subj: Roger, wait for a while.
Text:
Miki,
Roger, I'll immediately call him up and get an info for your "Main Frame." I am going to put that info during next orbit. Saya /\Z

3.4 The "K" Command

K <file#1>, <file#2>, <file#3> . . . <file#7>, <file#8>
K = KILL! Delete file(s) (messages) specified by file numbers. The <file#> is the same one described in R command. Up to eight files can be specified in a command line. A user can only delete files addressed solely to himself (ie, not to multiple users) or files he posted. (The W0RLI/WA7MBL equivalent command is also "K", except that multiple files can be killed at one time.)

3.5 The "H" Command

H = HELP! Entering H <cmd> gives additional information on that command. Entering only H will give a list of all available commands.

3.6 The "M" Command

M = Mine. List the latest 10 files (messages) that are either to or from the current user. Additional M commands list additional active messages. This command will be useful to save channel time when the user only wants to see his messages. (The W0RLI/WA7MBL equivalent command is "LM".)
JAS>M

NO.	DATE	FROM	TO	SUBJECT
111	10/10	G3AAJ	*	Harry in London
103	10/06	JR1FIG	JA9BOH	Uchiawase wa raishuu?
102	10/09	N7FDA	*	RS-232C card for PC-1089
100	10/08	JR1ING	JR1FIG	Sara ni kogata no TNC
95	10/08	N5AHD	JR1FIG	Automatic tracking system

Amateur Satellite Communications

Conducted By
Vern "Rip" Riportella, WA2LQQ
PO Box 177, Warwick, NY 10990

New Russian Satellite Sparks Surge of Interest

Every once in a while, we get to share the great joy of celebrating a rare and wonderful event in Amateur Radio: The birth of a new OSCAR.¹

On June 23, our Russian colleagues in space successfully launched a new OSCAR. RS-10 and RS-11 are alive and well, some 620 miles aloft.

Unlike earlier RS satellites, which had gone up on a single launcher containing as many as six separate satellites,² COSMOS 1861 was a single spacecraft carrying two Amateur Radio transponders and one special navigation transponder. Here's how the story unfolded.

The Radio Moscow announcement said COSMOS 1861 had been launched earlier in the day. That was Tuesday, June 23. The announcement said COSMOS 1861 carried Amateur Radio communications relay equipment in addition to its primary scientific and communications research payload. The new RSs were aloft at last! Within hours, G3IOR had his first access and QSO, confirming that the new birds were up and running. Soon, W0CY was also reporting access and initial tracking information.

According to the best current information, RS-10 and RS-11 are identical except for operational frequencies. Each RS apparently uses *three distinct bands* in various combinations to achieve *five distinct modes* of operation in addition to its auxiliary Robot repeaters. (A Robot is an automatic QSO machine that will engage you in a CW QSO when addressed properly.)³

On each RS, 15 meters is used exclusively as an uplink band, 10 meters is used exclusively as a downlink band and 2 meters can be employed as either an uplink or downlink band. The overall frequency scheme is delineated in the accompanying table.

The desired orbit was attained precisely. The nodal period is 105.0245 minutes; the orbital increment is 26.3824 degrees west per orbit. A reference orbit for Sunday, July 5, is 00:14:31 at 61.2 degrees West. Average height is close to 1000 km (621 miles).

Next month, we'll explore the fascinating RS-10 and RS-11 telemetry suite. Later, we'll return to the techno-sport theme begun last spring.

RS-10 and RS-11 Operating Frequencies

Summary for both RS-10 and RS-11:

Mode A	2 meters up and 10 meters down.
Mode K	15 meters up and 10 meters down.
Mode T	15 meters up, 2 meters down.
Mode KT	15 and 10 meters up, 2 meters down.
Mode KA	15 and 2 meters up, 10 meters down.

Beacons can carry telemetry or Robot downlink.

RS-10

(Note: All uplink and downlink frequencies are given in MHz.)

Mode A: 145.860-145.900 up yields 29.360-29.400 down.

Uplink		Downlink	
		29.357	beacon
145.860		29.360	passband limit, lower
145.870		29.370	
145.880		29.380	passband center
145.890		29.390	
145.900		29.400	passband limit, upper
		29.403	beacon

Robot A:

Uplink	Downlink
145.820	29.357 or 29.403

Mode K: 21.160-21.200 up yields 29.360-29.400 down.

Uplink		Downlink	
		29.357	beacon
21.160		29.360	passband limit, lower
21.170		29.370	
21.180		29.380	passband center
21.190		29.390	
21.200		29.400	passband limit, upper
		29.403	beacon

Robot K:

Uplink	Downlink
21.120	29.357 or 29.403

Mode T: 21.160-21.200 up yields 145.860-145.900 down.

Uplink		Downlink	
		145.857	beacon
21.160		145.860	passband limit, lower
21.170		145.870	
21.180		145.880	passband center
21.190		145.890	
21.200		145.900	passband limit, upper
		145.903	beacon

Robot T:

Uplink	Downlink
21.120	145.857 or 145.903

Mode KT: 21.160-21.200 up yields 29.360-29.400 and 145.860-145.900 down.

KT Uplink	K Downlink		T Downlink	
			145.857	beacon
	29.357	beacon	145.860	passband limit, lower
21.160	29.360	passband limit, lower	145.870	
21.170	29.370		145.880	passband center
21.180	29.380	passband center	145.890	
21.190	29.390		145.900	passband limit, upper
21.200	29.400	passband limit, upper	145.903	beacon
	29.403	beacon		

Mode KA: 21.160-21.200 up and 145.860-145.900 up yields 29.360-29.400 down.

K Uplink	A Uplink	KA Downlink	
		29.357	beacon
21.160	145.860	29.360	passband limit, lower
21.170	145.870	29.370	
21.180	145.880	29.380	passband center
21.190	145.890	29.390	
21.200	145.900	29.400	passband limit, upper
		29.403	beacon

RS-11

Mode A: 145.910-145.950 up yields 29.410-29.450 down.

Uplink		Downlink	
		29.407	beacon
145.910		29.410	passband limit, lower
145.920		29.420	
145.930		29.430	passband center
145.940		29.440	
145.950		29.450	passband limit, upper
		29.453	beacon

Robot A:

Uplink	Downlink
145.830	29.407 or 29.453

Mode K: 21.210-21.250 up yields 29.410-29.450 down.

Uplink		Downlink	
		29.407	beacon
21.210		29.410	passband limit, lower
21.220		29.420	
21.230		29.430	passband center
21.240		29.440	
21.250		29.450	passband limit, upper
		29.453	beacon

Robot K:

Uplink	Downlink
21.130	29.403 or 29.453

Mode T: 21.210-21.250 up yields 145.910-145.950 down.

Uplink		Downlink	
		145.907	beacon
21.210		145.910	passband limit, lower
21.220		145.920	
21.230		145.930	passband center
21.240		145.940	
21.250		145.950	passband limit, upper
		145.953	beacon

Robot T:

Uplink	Downlink
21.130	145.907 or 145.953

Mode KT: 21.210-21.250 up yields 29.410-29.450 and 145.910-145.950 down.

KT Uplink	K Downlink		T Downlink	
			145.907	beacon
	29.407	beacon	145.910	passband limit, lower
21.210	29.410	passband limit, lower	145.920	
21.220	29.420		145.930	passband center
21.230	29.430	passband center	145.940	
21.240	29.440		145.950	passband limit, upper
21.250	29.450	passband limit, upper	145.953	beacon
	29.453	beacon		

Mode KA: 21.210-21.250 up and 145.910-145.950 up yields 29.410-29.450 down.

K Uplink	A Uplink	KA Downlink	
		29.407	beacon
21.210	145.910	29.410	passband limit, lower
21.220	145.920	29.420	
21.230	145.930	29.430	passband center
21.240	145.940	29.440	
21.250	145.950	29.450	passband limit, upper
		29.453	beacon

Notes

¹I use the word "OSCAR" in a generic sense here.
²The last RS launch was on December 2, 1981. Six satellites, RS-3, 4, 5, 6, 7 and 8, were launched by a single rocket at that time.
³RS-5 and RS-7 both had Robots aboard, but both are probably defunct. To address a Robot you simply send: "RS-10 de (your call) AR" on the Robot channel. RS-11 has a Robot, too, so you would substitute its call sign in your addressing transmission. For example, RS11 DE WA2LQQ AR should work fine. The Robot will reply with a signal report and the QSO serial number. In the past, special certificates were sent to Robot users by the sponsoring organization in the USSR, thought to be DOSAAF.

Amateur Satellite Communications

Conducted By
Vern "Rip" Riportella, WA2LQQ
PO Box 177, Warwick, NY 10990

New Russian Satellite Sparks Interest Surge: Part 2

Last month, we looked at the two new Russian Amateur Radio transponders, RS-10 and RS-11.[1] This month, we'll look at the telemetry suite of RS-10 and RS-11 to see what it can tell us.

Telemetry means measurement from a distance. In the present context, telemetry indicates the situation in and around a satellite. Through careful monitoring of the telemetry, one obtains an interesting and informative appreciation of the goings-on within the spacecraft. There's a lot of enjoyment and education that can be derived from simply monitoring them on their easily heard beacon frequencies.[2,3]

The telemetry is sent in CW at 20 WPM. The information conveyed represents various status points and measurements. There are 16 status points and 16 measurements sent in a repeating cycle. The beginning and end of each cycle is indicated by the identifier "RS10" or "RS11." A complete cycle comprises the RS identifier, 16 "words" of telemetry, and the RS identifier. A complete cycle is called a *telemetry frame*.

Each of the 16 words is comprised of two parts: an *alpha* part and a *numeric* part. Both parts are two characters long. For example, take the telemetry word "IG45." "IG" is the alpha part, and "45" is the numeric part. From Table 1, we see that "IG" indicates channel 4 information is being sent.[4] The alpha portion of channel 4 codes the status of the 15-meter uplink receiver.

The first character in the alpha portion codes the state of the binary variable being sent (refer to Table 3, next page). The "I" in the "IG45" example indicates the 15-meter receiver is off. Had the first character been an "N," we would have known the 15-meter receiver was on.

The numeric part of each telemetry word gives a value for a parameter of interest. For example, the temperature of the 10-meter transmitter is coded in channel 9. The value of the numeric portion of the telemetry word can vary from 00 to 99. To convert the numeric value to meaningful units of current, voltage or temperature, process the numeric portion of each word through the equation for that word. The equations are given in Table 3.

Here's a complete analysis of our example, "IG45." The "IG" indicates the status of channel 4: 15-meter receiver on/off. The "I" means it's off. Taking the "45" portion of our example and dropping it into the equation for channel 4, we get AGC voltage, V = n/5 = 45/5 = 9 V.

When a command station is accessing the system, the telemetry format is modified slightly. The modification takes the form of an extra dot or dash attached as the first Morse element of the first alpha character of each word sent (see Table 2).

Logging trends in telemetry is one of the most fascinating and edifying aspects of OSCAR work. It especially prepares young minds for the discipline, organization and record keeping essential to engineers and scientists. Next month, I'll detail how you can try out the fascinating Robot QSO machines on RS-10 and RS-11.

Notes

[1] Both transponders ride along with the primary payload as a navigation system for ground users) on a "bus" called Cosmos 1861. All were launched from the Soviet Union on June 23, 1987.

[2] Many secondary schools have included telemetry analysis of OSCAR satellites, together with orbital analysis/prediction, in their science curricula. *The Satellite Experimenter's Handbook* (SEH), published by ARRL and written by Dr Martin Davidoff, K2UBC, was derived from an earlier book by Davidoff, *Using Satellites In the Classroom*. SEH is a good place to start for ideas on including telemetry analysis and tracking topics in the scholastic classroom.

[3] Beacons are (RS-10) 29.357, 29.403, 145.857, 145.903 MHz; (RS-11) 29.407, 29.453, 145.907, 145.953 MHz. A complete frequency list appeared in October 1987 *QST*.

[4] Similarly, NG would indicate channel 4 is being sent.

[5] Total receiver attenuation implemented is sum of channels 2 and 3. Attenuation can take values of 0, 10, 20 or 30 dB.

[6] AMSAT NA is the world's largest organization of Amateur Radio satellite operators. Free information is available for an SASE (or SAE with IRCs) sent to AMSAT, PO Box 27, Washington, DC 20044, tel 301-589-6062.

Table 1
Summary Telemetry Matrix

General telemetry word form is: XY## where X is I, N, A or M (or may be modified as shown in Table 2); Y is a channel designator and can be S, R, D, G, U, W, K or O. ## is a two-digit number. Details are given in Table 3.

	X	Y	Meaning of XY	Parameter "##" Indicates
1	I/N	S	Telemetry sampling period	Power supply voltage over sample period
2	I/N	R	RX attenuation	2-m TX output power
3	I/N	D	RX attenuation	10-m TX output power
4	I/N	G	15-m RX status	15-m RX AGC voltage
5	I/N	U	2-m RX status	2-m RX AGC voltage
6	I/N	W	Command station channel	Command station AGC voltage
7	I/N	K	10-m service command	10-m beacon power output
8	I/N	O	2-m service command	2-m beacon power output
9	A/M	S	Status of 1st memory board	10-m TX temperature
10	A/M	R	Status of 2nd memory board	2-m TX temperature
11	A/M	D	Memory loading channel	20-V power supply temperature
12	A/M	G	Code Store memory status	9-V power supply temperature
13	A/M	U	Memory dump via channel	Control of backup 9-V power supply
14	A/M	W	15-m Robot RX status	IF voltage of 15-m Robot RX
15	A/M	K	2-m Robot RX atten.	IF voltage of 2-m Robot RX
16	A/M	O	Robot QSO counter where 00-32 QSOs logged is indicated as 00 and 33-128 QSOs is indicated by values in the range of 80-99.	Command channel 2-m power output

Table 2
Format Modification During Command Station Access

Channel Number	X with No Command Station Access	X with Command Access Via 15 Meters (Add leading dot)	X with Command Access Via 2 Meters (Add leading dash)
1-7	I or N	S or R	D or G
8-16	A or M	U or W	K or O

Explanation: The X character in the general telemetry word XY## normally indicates just a binary state, such as on or off. By substituting a different character set, the X character can indicate more. The first character is changed to indicate a command station is accessing the satellite through command channel. When X = I, N, A or M, the command station is *not* accessing the satellite. When X = S, R, U or W, the command station is accessing via the 15-m uplink. When X = D, G, K or O, the command station is accessing via the 2-m uplink. Thus, there are three character sets for use in the first, or X, slot. The first has two code elements; the other adds a leading dot or dash to make the new character sets, which use *three* code elements for each character.

Table 3
Detailed RS-10/RS-11 Telemetry Analysis

Each channel is sent as a four-character word, eg, XY##, where XY is a two-character alpha indicator, ## is a two-character numeric value.

Ch No.	"X" and its meaning	"Y" Meaning of "XY" part
1	I = 90 minutes N = 10 minutes	S Telemetry sampling period. ## = Power supply voltage over sample period where V = n/4 volts. Example: IS80. Sample period = 90 minutes; battery voltage = 20
2	I = 20-dB attenuator N = 0-dB attenuator	R RX attenuation. [See note 5] ## = Power output of 2-m TX where W = n/10 in watts Example: NR25. 0 dB attenuation; power = 2.5 W
3	I = 10-dB attenuator N = 0-dB attenuator	D RX attenuation. [See note 5] ## = Power output of 10-m TX where W = n/10 in watts Example: ID19. 10 dB attenuation; power = 1.9 W
4	I = Off N = On	G 15-m RX status.
5	I = Off N = On	U 2-m RX status.
6	I = Off N = On	W Special command station channel
7	I = 1000 mW N = 300 mW	K 10-m beacon power output ## = Special command station AGC voltage, where V = n/5 in volts Example: IW00. Command channel off; AGC voltage = 0
8	I = 1000 mW N = 300 mW	O 2-m beacon power output ## = Service command parameter, 10-m mode command mode off Example: IK00. 10-m beacon 1000 mW out;
9	A = Off M = On	S Status of 1st memory board ## = Service command parameter, 2-m mode command mode off Example: IO00. 2-m beacon 1000 mW out; command mode off ## = 10-m TX temperature, where T = n – 10 in °C Example: AS35. 1st memory board off; temp of 10-m TX = 25°C

Ch No.	"X" and its meaning	"Y" Meaning of "XY" part
10	A = Off M = On	R Status of 2nd memory board ## = 2-m TX temperature, where T = n – 10 in °C Example: AR23. 2nd memory board off; temp of 2-m TX = 13°C
11	A = Open M = Closed	D 1st memory board ## = 20-V power supply temperature, where T = n – 10 in °C Example: AD38. Memory channel open; 20-V supply temp = 28°C
12	A = Open M = Closed	G Code store memory status ## = 9-V power supply temperature, where T = n – 10 in °C Example: MG31. Code store memory closed; 9-V supply temp = 21°C
13	A = 10 m M = 2 m	U Memory dump via channel ## = Control parameter backup 9-V power supply, where V = n/5 volts Example: AU00. Memory dump via 10 m; backup 9-V supply control off
14	A = 10 dB M = 0 dB	W Attenuator setting of 15-m Robot RX ## = IF voltage of 15-m Robot RX, where V = n/5 in volts Example: AW46. 15-m Robot RX attenuated 10 dB; Robot IF voltage = 9.2
15	A = 10 dB M = 0 dB	K Attenuator setting of 2-m Robot RX ## = IF voltage of 2-m Robot RX, where V = n/5 in volts Example: AK46. 2-m Robot RX attenuated 10 dB; Robot IF voltage = 9.2
16	A = 1000 mW M = 300 mW	O Special command channel 2-m power output ## = Robot QSO counter where 00-32 QSOs logged is indicated as 00, and 33-128 QSOs is indicated by values in the range of 80-99. Example: AO89. Command channel, 1000 mW out; Robot QSOs counted, 33 < n < 128

Amateur Satellite Communications

Conducted By
Vern "Rip" Riportella, WA2LQQ
PO Box 177, Warwick, NY 10990

New Russian Satellite Sparks Interest Surge: Part 3

In previous months, I have discussed the new Russian Amateur Radio transponders, RS-10 and RS-11, and their telemetry suite. This month you'll learn how to use the Robot QSO machines on these Russian birds.

A Robot, as the term is applied to the RS (Radio Sputnik) satellites, is an "auto-responder," or automatic QSO machine. In other words, it's a small computer on board the satellite that can engage you in a simple CW QSO, respond to your call, log it, and send you a radio QSL. Later, if the Soviet operators keep to the practice established years ago with the RS-5 and RS-7 Robots, you may get a QSL card in the mail, too.

As I explained in earlier columns, both RS-10 and RS-11 are on the same spacecraft, and have similar functions. However, they operate on different frequencies and cannot be in operation simultaneously. The frequencies that the Robots use are given in Table 1.[1]

An experienced satellite operator, Ray Soifer, W2RS, explains the Robot access procedure this way: First, listen for the Robot on one of the downlinks indicated in Table 1. If the Robot is ready for action, it will tell you so by sending CQ DE RS10 or CQ DE RS11 and then indicate the frequency on which it will listen for your call. Then, all you need to do is call the Robot as follows, using your call in place of W2RS, in the following example:

RS10 DE W2RS AR

If the Robot has heard your signal, it will respond with something like this:

W2RS DE RS10 QSL NR 775 OP
Robot TU USW QSO 775 73 SK

It is not clear what "USW" means, but the rest should be evident to most amateurs. TU is Thank You, obviously. QSLs may be sent to the usual QSL address, ie, Radio Sport Federation, PO Box 88, Moscow, USSR.

The RS birds built and launched by Russian Amateurs have traditionally been simple, rugged devices designed primarily to serve Soviet-bloc nations, where modern VHF and UHF radio equipment is difficult to get and very expensive. Consequently, RS satellites have tended to exploit the high frequencies (HF) to a much greater extent than Western satellites have in the last decade.

This may change somewhat in the future as a result of several factors. For one, the next few years will see a rise in solar activity. That will result in increased ionospheric density. Of course, this is a boon to HF users of F2 propagation. But have you considered what it might mean to a satellite emitting, say, 15-meter signals on the outside of the F2 layer?

Obviously, at the peak of the solar cycle, which will arrive in a few years, very little 15-meter energy will get through from satellites radiating from above the F2 layer. Indeed, during the last solar activity peak in the late 70s and early 80s, it was fascinating to listen to the 10-meter downlinks of various satellites such as RS-1, RS-2 and AO-8 play peekaboo through various holes in the ionosphere.

So, for their next generation of RSs, which will operate during the next peak of the solar cycle, our Soviet colleagues will likely plan more extensive use of VHF and UHF frequencies than they have in the past. This effort may be abetted by new cooperation that now seems possible between the RS builders and the builders of the OSCARs in the West.

I recently had the pleasure of meeting with Leonid Labutin, UA3CR, at the "Space Future Forum," in Moscow, and later at his home. Leo is widely recognized as the foremost Amateur satellite enthusiast in the Soviet Union. He has undertaken to form a new affiliate of AMSAT, AMSAT-UA, for the express purpose of facilitating joint spacecraft projects. This could, in turn, lead to more extensive use of VHF and UHF by RS builders and increase launch access for all the Amateur Satellite community.[2] The fruits of this new era of cooperation may begin to become evident by the time you read this column.

In general, cooperation in building and launching Amateur Radio satellites can have strong symbolic meaning as well as substantive benefits for Amateurs worldwide. Next month I'll begin to introduce you to the new AMSAT Phase 3C satellite due for launch in a couple of months.

Table 1
RS-10 and RS-11 Auto-Responder (Robot) Frequencies

Transponder	Mode	Uplink Frequency, MHz	Downlink Frequency, MHz
RS-10	A	145.820	29.357 or 29.403
RS-10	K	21.120	29.357 or 29.403
RS-10	T	21.120	145.857 or 145.903
RS-11	A	145.830	29.407 or 29.453
RS-11	K	21.130	29.407 or 29.453
RS-11	T	21.130	145.907 or 145.953

Leonid Labutin, UA3CR (left) and your column conductor met in Moscow last October to discuss joint Amateur Radio satellite projects.

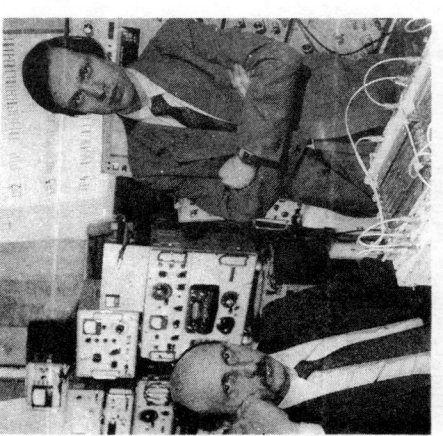

Alexandr Papkov (left) and Viktor Samkov, of the Tsiolkovskiy Institute near Kaluga, are shown with the RS-10 and RS-11 transponders. (Reprinted from the cover of May 1987 *Radio*.)

Notes

[1] A complete list of frequencies employed by the RS-10 and RS-11 transponders was published in this column in October 1987.

[2] Launches are the rarest of all resources in the Amateur Satellite community. Launches are so rare that when one is identified and committed to, it can—and does—change the entire Amateur Satellite community. AMSAT and its affiliates try to obtain launches from all possible sources. We will accept a launch from virtually any dependable source. Against the existing background of a supreme appetite for launches (commercial, military, amateur), recognize that the USSR currently out-launches the rest of the world combined by a 10 to 1 ratio! In this context, it certainly makes sense to seek a symbiotic relation with the RS team.

Amateur Satellite Communications

Conducted By
Vern "Rip" Riportella, WA2LQQ
PO Box 177, Warwick, NY 10990

The Future Is Up There

We now recognize the need for a broader perspective in designing future satellite systems. The critical design elements must include not just the spacecraft itself but the whole system, including the user interface and environment. Present and future users will require that next-generation satellites be more convenient and accessible.

Future systems will need to be easier to use. They should provide a reasonable performance payback for the users' investment in time (to learn techniques) and equipment. The most satisfactory solution to realizing these and other important objectives is, simply, "up." Up in altitude, up in frequency and, consequently, up in cost.

Why up? The accompanying figures illustrate the advantages of higher altitude. AMSAT has under consideration two major options for the next-generation high-altitude satellites. One would be the so-called Phase 4 plan for geosynchronous satellites. Phase 4 would have perhaps two geosynchronous satellites operational by decade's end. The other major option is a further refinement of the current Phase 3 program using two or more satellites in a true Molniya orbit inclined 63 degrees is being closely scrutinized.

Let's look at the latter advanced Phase 3D concept first. It would have two or more true Molniya orbit satellites phased in such a way as to provide nearly continuous coverage for the Northern Hemisphere. The orbital period would be 12 hours. This is important to reduce tracking complexity. The satellites would appear in the same part of the sky on a regular, easily predicted basis. As one satellite passed apogee, its partner would be on the rise. Later, the second one would appear in the same sky location as its partner earlier was located. They would continue in this bola-like arrangement indefinitely. The prime objective of nearly continuous coverage would be realized, since at least one bird would always be in view and "on line." And minimal tracking and antenna steering would be required.

Satellites of a Phase 3D option would build on existing technology. Although some stationkeeping would be required, current techniques using magnetic torqueing might suffice.[1] Despun bearings and other very expensive components would then be unnecessary, and costs would be reduced substantially.

What about Phase 4? The view from a single geosynchronous satellite such as AMSAT might consider was pictured last month. The engineering challenges involved in a Phase 4 system are formidable. In the end, however, the deciding factor may be economic. The key technologies would have to be purchased rather than fabricated by in-house AMSAT.[2] This would certainly drive the cost of even a single Phase 4 bird over the one-megabuck threshold. Whether in a Phase 4 geosynchronous orbit

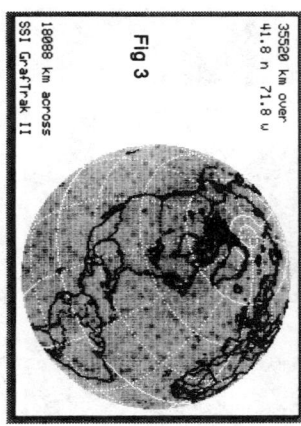

Fig 1
300 km over
41.8 n 72.8 u
3838 km across
SSi GrafTrak II

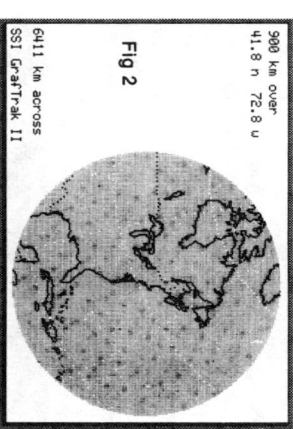

Fig 2
900 km over
41.8 n 72.8 u
6411 km across
SSi GrafTrak II

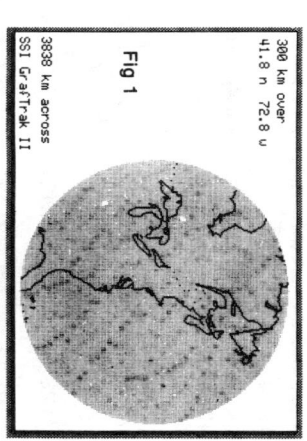

Fig 3
35520 km over
41.8 n 71.8 u
10000 km across
SSi GrafTrak II

One of the reasons for going "up" in altitude is coverage area. Fig 1 shows the view from a typical Space Shuttle mission if it were directly over Newington at its nominal flight altitude of about 300 km (186 statute miles). Coverage area extends north to Newfoundland, west to the Missouri River and south to Miami. Compare this to the coverage afforded by a typical low earth orbiter such as AMSAT-OSCAR 8, as shown in Fig 2, again positioned over Newington. Coverage is seen to extend north to central Greenland, west to the Rockies and south to Central America. Finally, compare with Fig 3 the view from a typical Phase 3-type elliptical orbit with apogee above Newington. Coverage now extends north over the pole to Asia, west to all of Europe and some of the Middle East and Western Africa, south to Argentina and Chile and west to Hawaii. Compare these three figures with the coverage from a typical geosynchronous satellite, as shown in the January 1986 edition of this column.

or a clever adaptation of a Phase 3 Molniya-type orbit, future satellites will surely embody our premise: The future is up.

But with increased coverage area comes the need for more transponder bandwidth. Clearly the 2-meter and 70-cm bands that now

support AO-10 traffic in a bit over 150 kHz cannot support the spectral requirements warranted by the coverage area of the next-generation satellites. The only answer is again "up." At the higher frequencies, 23 cm and up, lie the spectral resources needed for tomorrow's highly capable satellites.

Getting on the higher bands has never been easier, thanks to advances in key, gating technologies. Commercial manufacturers are about to unleash a torrent of equipment, making access to 23 cm and above easier than ever. The new ICOM 1271A all-mode 23-cm transceiver, I'm convinced, serves as harbinger of legions of UHF and even SHF equipment just around the corner. It will soon be possible to bolt together a complete 23-cm station. Soon, 13 cm will yield to the same market pressures that now have brought 23 cm to the brink of popularity.

The challenge of shaping the future of Amateur Radio satellites and, to an increasing degree, the course of the overall hobby is now engaged. Whether it be Phase 4A or Phase 3D, the outcome of current planning will determine the space-communications resources available to hams for at least the first half of the next decade.

Next month, we'll look at a few functions future satellites might provide, such as bulletin delivery to users through gateway repeaters.[3] Meanwhile, there's no time like the present to get your own station on OSCAR. Why wait for the future? Meet it halfway![4-6]

Notes

[1] By contrast, magnetic torqueing as a means of stationkeeping is not possible at geosynchronous altitudes because the geomagnetic field is too weak at 22,300 miles.

[2] The key technologies involve either three-axis stabilization or despun bearings. AMSAT can expect little help on either account from donations in kind and has very little likelihood of being able to develop this hardware internally.

[3] Project Linkup will demonstrate bulletin delivery to repeaters using AO-10 Mode L and FM for high-quality audio. Regular AMSAT and space-interest-oriented bulletins will be available to all who interface their stations and/or repeaters. Free info is available from the author (at the address at the top of this column) for interested individuals and groups. An SASE is required, please. First on-the-air tests are scheduled for first-quarter 1986. Project NATCOM will provide nationwide AMSAT net connectivity for local interest groups as well as AMSAT-affiliated clubs and repeaters using a telephone bridge provided by the Darome Connection, a commercial telephone conference service in Minneapolis. Free NATCOM information is also available for an SASE to this column conductor.

[4] Help in getting started can be obtained from AMSAT Area Coordinators. The name and address of one in your state or province can be obtained for an SASE to this column conductor.

[5] Information on the GRAFTRAK II software, used to produce the figures as well as other software available in the AMSAT Software Exchange, is available in the new ASE 1986 catalog, free for an SASE to this column conductor.

[6] Users of the ARRL OSCARLOCATOR can obtain current AO-10 ground-track templates for an SASE to this column conductor.

Amateur Satellite Communications

Conducted By
Vern "Rip" Riportella, WA2LQQ
PO Box 177, Warwick, NY 10990

NEXT Generation Satellites on the Horizon

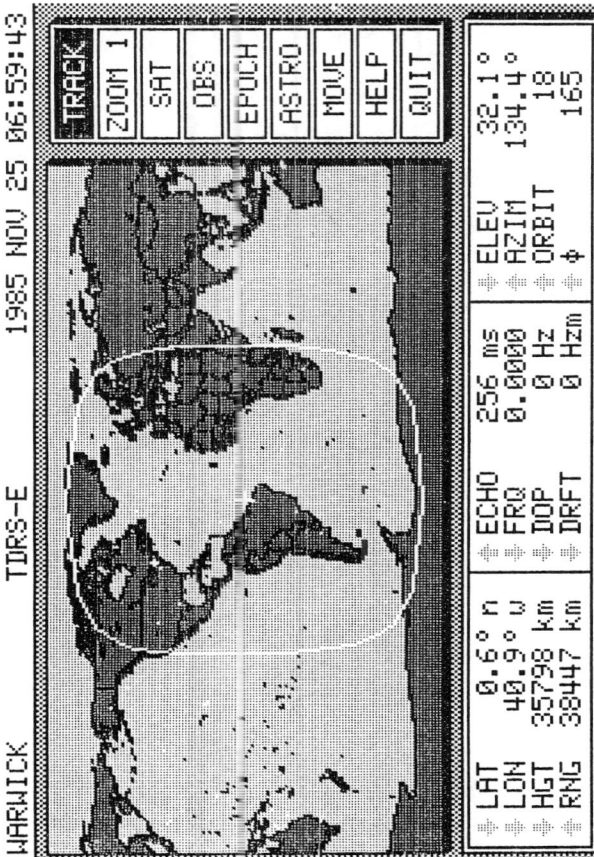

Geosynchronous satellite coverage is typified by this plot of the footprint of TDRS East. It is positioned just north of the coast of Brazil. Areas within the curve can intercommunicate via the satellite.

Despite the enjoyment derived in tracking satellites, there are circumstances when it's inconvenient to wait for OSCAR to appear. To the beginner, moreover, it's unclear why the satellite need move at all. "Hummingbirds and balloonists remain motionless, don't they?"

A satellite needs motion to stay in orbit. Without a velocity component away from earth, it would plummet.[1] When a satellite's motion away from earth and gravity are in balance, the trajectory closes on itself to become an "orbit."[2] The higher the satellite, the less velocity it need have to remain in orbit.[3] Today's "stationary" geosynchronous satellites are in orbits 22,300 miles over the equator. But even though they appear stationary, they too have a significant velocity component directed away from earth.

After two decades of experimenting in lower orbits, AMSAT is now charting new conceptual grounds for the next-generation OSCARs. Will they be Phase 4 geosynchronous satellites? What could a Phase 4 satellite system do?

You're motoring down the San Bernardino Freeway. You pick up your radio, punch in a few numbers and instantly you're in contact with your cousin, who's also motoring down the highway. Not really exciting, you say. You already do that? Did I mention that your cousin was motoring down the Trans-Canada highway west of Montreal? A satellite gateway in the San Gabriel mountains and another in the Gatineaus has linked these two motorists through an AMSAT geosynchronous satellite high over Ecuador.

You're sitting at your computer whacking out a reply to an electronic-mail QSL card. You hit the return key and a "MSG DELVD" advisory appears on your screen. The QSL card now resides in an accountable processor in Tucson. When the addressee, a Boston amateur, comes on frequency, the QSL will be forwarded automatically. The 9.6-kbit/s channel is provided by AMSAT's AMSTAR placed high over the Andes.

A severe earthquake hits coastal Mexico. Normal communications are obliterated. Soon a portable AMSTAR terminal at the site is on the air and a high-quality channel is available from the disaster site to every capital in the Western Hemisphere. Voice and packet data flow continuously for days.

Sitting in the canoe you avoid making any sharp movements. The lake is calm; anchors are positioned to hold the canoe steady. The small dish clamped to the bow rail is pointed right at Orion's belt where, you figured out earlier, AMSTAR would be. Your digital watch beeps in the dark to announce the hour. Donning your headset, the cosmic noise of innumerable stars supplants the din of an apparently equal number of katydids. As you fiddle with the squelch, your lake transforms to a tranquil, 75-acre hamshack. The voice in your ear announces, "Here are tonight's Amateur Radio bulletins via AMSTAR."

These four examples typify system capabilities being considered for the next generation of OSCARs. But to get there will require a much larger support base—a base comprising a larger portion of the Amateur Radio community than is now involved.

The motions of OSCARs are determined by physics. In confrontation with Newton and Kepler, user convenience is subordinate. For amateur satellites to reach their full service potential, however, they must serve the much larger community which, surveys indicate, places a premium on access convenience. Furthermore, while those currently on OSCAR often have an abiding interest in the medium itself, i.e., OSCAR, many others would prefer to ignore the medium entirely and get on to the matter of communicating. They seem to be saying, "One needn't get into the clockworks to be interested in the time of day!"

In response, AMSAT is engaging in some healthy introspection. Many are coming to feel what is needed now is a "utility satellite" system. In this line of thinking, advanced capabilities are provided through simple, convenient access enabled by superior engineering. In the months to come, AMSAT will weigh concepts for the next generation of satellites. It will study tentative goals for spacecraft design, user capability, financial resources, and others.

It is timely for current OSCAR user and potential user suggestions. Those experienced on OSCAR have a special perspective since they've conquered the beast; done it the hard way. But those who've never worked OSCAR also have something to say here. It would be helpful to have your ideas, too. How would you like OSCAR to perform, and how much would you be willing to put into your current station to achieve that new capability?

Amateur Radio will have geosynchronous satellites sooner or later. Given their druthers, many would have it sooner rather than later! If you would like to participate in the next-generation OSCAR planning, we'd be glad to hear from you. Meanwhile, AMSAT's Project Linkup will demonstrate the feasibility of bulletin delivery both directly and through gateway repeaters. Concept validation will lead to possible regular service before the end of 1986 using current-generation Phase 3 satellites. Want to be a part? An SASE will bring you information on getting started now. Avoid the last minute rush!

Next time we'll discuss potential Phase 4 system capabilities in detail.

Notes

[1] Similarly, a hummingbird *sans* aerodynamic lift and a balloonist without buoyancy would plummet.

[2] On the other hand, some situations result in starkly different trajectories. A particularly rude one on the low end of the velocity scale ends with a rather direct path to Mother Earth—reentry the hard way ... at Mach 25. At the other end of the scale, we find trajectories that are described by unique S parameters: Space Sayonara Syndrome, aka escape velocity.

[3] Very low satellites, i.e., about 100 miles, orbit in about 1.5 hours. At 900 miles, orbital periods are about 2 hours. At 22,300 miles, geosynchronous altitude, the orbital period is 24 hours. That's why it's "geosynchronous." At 230,000 miles, the moon's orbital period is about 28 days.

The ARRL Satellite Anthology
PROOF OF PURCHASE

ARRL MEMBERS

This proof of purchase may be used as a $.50 credit on your next ARRL purchase or renewal 1 credit per member. Validate by entering your membership number — the first 7 digits on your QST label — below:

FEEDBACK

Please use this form to give us your comments on this book and what you'd like to see in future editions.

Name _____ Call sign _____

Address _____ Daytime Phone ( ) _____

City _____ State/Province _____ ZIP/Postal Code _____

From _____

Editor, The ARRL Satellite Anthology
American Radio Relay League
225 Main Street
Newington, CT USA 06111

······ please fold and tape ······

Please affix
postage. Post
Office will not
deliver without
sufficient postage.